John Milton Scudder

On the Use of Medicated Inhalations in the Treatment of Diseases

of the Respiratory Organs

John Milton Scudder

On the Use of Medicated Inhalations in the Treatment of Diseases of the Respiratory Organs

ISBN/EAN: 9783337812324

Printed in Europe, USA, Canada, Australia, Japan

Cover: Foto ©berggeist007 / pixelio.de

More available books at **www.hansebooks.com**

ON

THE USE

OF

MEDICATED INHALATIONS

IN THE TREATMENT OF

DISEASES OF THE RESPIRATORY ORGANS,

BY

JOHN M. SCUDDER. M. D.

AUTHOR OF "A TREATISE ON THE DISEASES OF WOMEN," "THE ECLECTIC PRACTICE OF MEDICINE," "THE ECLECTIC MATERIA MEDICA AND THERAPEUTICS." LATE PROFESSOR OF THE PRINCIPLES AND PRACTICE OF MEDICINE IN THE ECLECTIC MEDICAL INSTITUTE, CINCINNATI, OHIO, ETC., ETC.

WITH AN APPENDIX ON

DISEASES OF THE NOSE AND THROAT,

BY

WM. BYRD SCUDDER, M. D.

PROFESSOR OF OPHTHALMOLOGY AND OTOLOGY AND DISEASES OF THE NOSE AND THROAT IN THE ECLECTIC MEDICAL INSTITUTE, CINCINNATI, OHIO.

FOURTH EDITION.

Preface to the Fourth Edition.

During the past eight years there has been considerable demand for a new edition of Professor J. M. Scudder's work on "Medicated Inhalations."

For various reasons it has been thought best not to alter the text in the edition of 1874, embraced in pages 1 to 91 of the present volume, which, it must be remembered, was written during the infancy of the study of Inhalations.

The new part of this work, which is added as an Appendix, pages 91 to 147, treats more particularly of *Diseases of the Nose and Throat*, and is in a field that has not been covered by eclectic literature. The price of this work as enlarged, has been reduced from $2.00 to $1.00.

The study of this work may stimulate and add an impetus in this line among our practitioners, and if so, the labor of the undersigned will have been amply repaid.

Wm. Byrd Scudder, M. D.

Cincinnati, May, 1895.

PREFACE.

The subject of direct medication in diseases of the air passages is beginning to attract the attention of physicians, as a field likely to be worked with advantage. It would doubtless have received more attention, but for the fact that it was made a specialty by ignorant persons, and extensively advertised as effecting cures, where the common means were of no avail.

It is evident that much good might result from the judicious use of local applications to diseased surfaces, if these could be used with anything like the degree of definiteness that attends their employment elsewhere. And it is the object of this monograph, to show that this can be done with the improved apparatus now employed.

As the study of such a subject necessarily requires years, and the united effort of a number of observers, in order to develop it, I need not offer any apology for such imperfections as may be discovered in the present work. It is written to draw more attention to the subject, and to present the practitioner with the results of three years study, as a basis for the commencement of the practice.

The author wishes to be distinctly understood, as claiming for inhalations only the position of adjunct means in many affections, and a principal means in but one or two. Insisting that it must be combined with a judicious general treatment if we desire to obtain the best results. But with such treatment, it will many times relieve certain symptoms, and cure certain structural diseases, that stand in the way of the patient's recovery.

As Dr. MADDOCK well remarks: "No one deprecates quackery in every shape and form more than ourselves, but we must say, that when the generality of practitioners are confessedly incapable of combating the ravages of pulmonary diseases, it is neither consistent with reason nor humanity, to expect that the public will stand supinely by, and see the inroads of disease unchecked or unalleviated, and leave their friends or relatives a prey to misery and despair, without making one effort to arrest its progress or cheer the mind by giving trial to new remedies or modes of treatment, especially if supported by reputable testimony. Narrow-headed men, with narrower minds, may whisper obstacles and superciliously condemn a new proposal, and pronounce it useless, without inquiry or any opportunity of seeing or judging of what they advise others to reject: but the abuse or discountenance of such members of the profession can avail but little, and only proves that such persons must be contracted not only in their ideas. but regardless of the advantages of their patients or the advancement of their profession. The realities, however, of inhalation are now by far too numerous and too well attested to be put down by contumely, or even by indifference ; and we doubt not but that from the number of authentic instances, which we almost daily receive, of its applicability to the

treatment of pectoral diseases, that such an amount of evidence will soon have been amassed, as to shake the belief of
the most inveterately sceptical, for in the investigation of
truth the illustration of its principles are never insulated;
and however one manifestation of its presence may be kept
from receiving general acceptance by the efforts of prejudice,
others will present themselves with a constancy which shall
render their distinctive character as incontrovertible as their
real existence."

As an introduction to the subject, the little work of W.
Abotts Smith, M. D., of London, is republished. No better
argument could be employed to recommend inhalation as a
means of cure, and to stimulate research in this direction.

98 *West Sixth Street, Cincinnati.*

I.

ON THE INHALATION OF GASES AND

MEDICATED VAPORS.

BY W. ABBOTTS SMITH, M. D., M. R. C. P.

So much has recently been said and written in extra-professional circles on the subject of inhalation as a remedial agent, and so evident is the fact, that many persons look upon it as a novel mode of treatment, that it will not be altogether uninteresting to commence this general description of inhalation by showing the opinions of some of the most ancient Professors of Medicine concerning it, in order to prove that it possesses high claims to antiquity.

In their descriptions of the treatment of catarrh, coryza, and cough, Galen, Aetius, Rhases, Ægineta, and other ancient authorities recommend the inhalation of the fumes of various substances, which are ordered to be ignited, so that the person affected may draw the vapor arising from them into the throat and lungs through a funnel. Haly Abbas and other celebrated Arabian physicians, who, at a later period, were the sole possessors of medical learning, also speak favor-

ably, amongst other remedies, of the inhalation of the
vapors of vinegar, camphor, and other substances   In
asthma and phthisis, again, similar fumigations were
recommended; and we find one old writer, Avicenna,
speaking in high terms of the inhalation of the vapour
of pine fruit, a plan which has recently come into vogue
in some parts of Germany.   References may also be met
with in some of the older treatises on Medicine to
fumigations with certain mineral substances, reduced to
the condition of vapor, in the treatment of secondary
syphilis.

It would be easy to multiply these instances of the
value which was set upon inhalation, not only by the
older medical writers, but also by those who have lived
in later times.   Yet, with all this, there are many who
would have others believe that inhalation is a new
method of treatment, and who, through combined igno-
rance, avarice, and love of notoriety, hold out specious
hopes of recovery to persons whose cases are unfortunately beyond human aid.   So much, indeed, for the
claim of novelty which has been put forward in respect
to inhalation.

Still, there is much in the practice of inhalation when
rightly used, to make it worthy of employment in many
cases of pectoral disease ; and I believe that most
medical men will agree with me that it is the indiscriminate use of inhalation, and not a judicious resort
to it, when it is suitably indicated, that is to be depre-
cated.*

* I have heard the opinion advanced, but as I venture to think, very
illogically, that because inhalation has been unfortunately adopted by
empirics, this method of treatment should be discountenanced by the
Medical Profession. I must confess I cannot see the force of this argument. If there be anything in inhalation, it is our duty, in the interests
of our patients, to give them the benefit of it, and to show to what extent

Like the Turkish bath, and many other agents of undoubted remedial value, inhalation occupies a lower position in the therapeutic scale than it would do were it not for the discredit which has been brought upon it by the extravagant views which have been advanced by its supporters, and which have led to consequent disappointment.

Inhalation was confined to fumigations with the vapors of various vegetable, and occasionally mineral, substances, until towards the close of the last century, when the discoveries which were made of oxygen, and other gases, induced several distinguished physicians and chemists, including Priestly, Beddoes, Cavendish, and Humphry Davy, to anticipate most favorable results from the inhalation of different gases in the treatment of consumption and other diseases. For the purpose of practically carrying out the views of the promoters of the system of Pneumatic Medicine, as it was called, an institution was founded at Bristol, by Dr. Beddoes, with Humphry Davy (then just beginning his illustrious career) as superintendent. Capacious reservoirs were constructed for the reception of large quantities of oxygen, carbonic acid, carburetted hydrogen, and other gases, and patients flocked thither in considerable numbers from all parts of the country to avail themselves of Dr. Beddoes' treatment. Various circumstances, amongst which may be comprised the too sanguine ideas of the founders of this establishment, and the costly nature of the apparatus employed, led to

it may be rendered useful in the alleviation or cure of disease; but to relinquish any remedy because some ignorant person has laid hands upon it, and from his want of knowledge, or some other cause, has pretended to discovered merits in it which it does not posess, seems to be opposed to common sense

the ultimate abandonment of the project, but not before
researches had been conducted on a sufficiently extensive scale to show that much good might be derived
from the inhalation of some of the gases employed,
especially of oxygen, if some plan could be devised by
which they could be generated in a less costly and
more portable manner, so that patients could use them
at their own homes, instead of their being compelled
to undertake long and hurtful journeys. Those of our
readers who desire additional details of the history of
Dr. Beddoes' "Medical Pneumatic Institution," will find
much interesting information in "Beddoes' Letter to
Erasmus Darwin, M.D., on a New Method of treating
Pulmonary consumption," published in 1793, in Davy's
"Chemical and Philosophical Researches,"* written by
that author whilst he was superintendent of the Medical Pneumatic Institution, and published in 1800, and
in the valuable Memoir of Sir Humphry Davy, edited
by his brother Dr. John Davy.

The impediments just referred to, thrown in the way
of administering oxygen by inhalation, appear to have
baffled later writers on pulmonary affections. Laennec,

* During his residence at Clifton, near Bristol, where the Medical Pneumatic Institution was situated, Davy experimented upon most of the gases
then known. Dr. Druitt, in his very useful "Surgeon's Vade Mecum"
(seventh edition, p. 702), points out the important fact that Davy, in con-
ducting some experimental researches on the properties of nitrous oxide,
found that its inhalation mitigated the pain of cutting a wisdom-tooth, and
from other circumstances connected with its inhalation, he threw out the
hint that as it appeared to be "capable of destroying physical pain, so it
might probably be used with advantage during surgical operations.'
From this discovery of Davy's and that of cotemporaneous observers
that the inhalation of ether, pure, or medicated with conium or some other
vegetable sedative, allayed the irritation in spasmodic asthma and
whooping-cough, may be traced the inestimable boon of Chloroform, as
an anæsthetic.

in his "Treatise on Mediate Auscultation, and Diseases of the Heart and Lungs," remarks that "no means could seem better calculated to combat the dyspnœa which arises from increased want of respiration, in spasmodic asthma, than the inhalation of pure oxygen," but, as he adds, the difficulty of promptly procuring it is a complete bar to its employment.

This difficulty no longer exists, for oxygen can now be generated in sufficient quantity, and at such a moderate cost, as to allow of its general use; and which is of still greater importance, the trials which have been made of it by various observers show that it possesses remedial powers of no small value.

The largest and most complete series of experiments with oxygen gas both upon animals and human beings, are those which, during several years past, have been conducted by my esteemed friend M. Demarquay, of Paris, in conjunction with M. Leconte. These careful and trustworthy observers have recently embodied the results of their researches in some reports presented to the Academy of Sciences of Paris; and their conclusions concerning the physiological and therapeutical properties of oxygen so closely coincide with those of other experimenters, that they may be advantageously epitomised in the following description of the effects of oxygen when inhaled.

The first series of experiments, which comprised a large number of observations, had relation to the action of oxygen on animals only.* In these experiments,

---

* In reference to experiments conducted upon animals, physiologists have of late been attacked, in no measured terms of reproach and abuse, for the alleged cruelties committed by them in the prosecution of their inquiries. Without entering upon a lengthy defense of scientific men against such charges, it may be observed, that a physiologist, working for the im-

it was ascertained that dogs are able to respire during
a long period, as much as thirty or forty litres of oxygen,
or even more (a litre, it may be remarked, being equal
to 1·76 English pints), without injury, the only apparent
result of the inhalation being to render the animal more
lively and to improve his appetite. In some instances
large wounds were made in the axillary region, and the
animals were made to respire oxygen when these
wounds were healing. It was then observed that they
became brightly injected with arterial blood, and that
a transparent, serous fluid was effused at their surface;
and, further, that as the inhalation of the gas was continued, numerous petechiæ, or ecchymoses, were produced. In order to ascertain whether similar results
would be produced by the injection of oxygen into the
venous system, a series of injections into the external
jugular vein were made, and the same effects were
produced as when inhalation was practiced. The principal results shown in the summary of the researches of
MM. Demarquay and Leconte into the effects of inhalation of oxygen upon animals are :— 1. That after
death, caused by the constant respiration of this gas,
the muscular system of the animals subjected to it was
found to be in a very turgescent condition. 2. That,
contrary to the opinion advanced by Broughton, an

provement of science, and the ultimate benefit of the human race, inflicts
less pain and torture in the whole course of a year than many a sportsman
would cause in a few days' shooting, indulged in merely for the gratifi-
cation of his cwn pleasure. A rabbit (one of the animals most commonly
selected for physiological experiments) would suffer far less pain at the
hands of a physiologist, who, whenever practicable, would most likely
render it insensibe by the administration of chloroform, previously to the
commencement of his experiments, and who would put it to a speedy death
afterwards, than the same animal would suffer at the hands of a sportsman,
if, with broken leg or mangled body, it had crept away into a thicket to
meet a tedious termination to its misery.

early inquirer into the physiological action of oxygen, the venous and arterial symptoms retained their natural color. 3. That, contrary to the assertion of Beddoes, no organ, however vascular it might be, was ever found to be the seat of either inflammation or gangrene; and 4. That the muscular system assumed a peculiar rosy hue.

MM. Demarquay and Leconte next proceed to describe the action of oxygen upon man. In the first place, when it is applied locally to the surface of wounds, by means of special caoutchouc apparatus it gives rise to a slight sense of heat and tingling, without pain. In the course of a few hours the suppuration becomes diminished in quantity and consistency, and the granulations assume a greyish color, and appear to be smaller in size. After the oxygen has been removed, they again become red and turgescent; and, if the application of the gas be renewed for several hours daily, for some succesive days, more or less inflammatory action is induced. One of the most remarkable effects of the oxygen, topically applied, is the rapid manner in which it modifies the conjestive redness commonly present at the circumference of a wound; and, in this manner, the redness surrounding ulcers of the extremities, and the injection of the skin remaining after eczema, may be readily removed. This property of oxygen, when applied locally, will, doubtless, at some future period, be turned to beneficial results in the treatment of indolent and other forms of ulcers and wounds, as also to that of certain affections of the skin.*

* While briefly refering to the original uses of oxygen, the fact may be here stated, that the tunica vaginalis has been injected with this gas without inconvenience, hydrocele having in one instance subsequently undergone cure

It is, however, with the effects of oxygen, when inhaled, that I purpose to deal in the present paper. I
must, therefore, pass on to this point.

Demarquay and Leconte, with their pupils and
friends, found that they could readily inspire a dose of
from twenty to thirty litres without any inconvenience
resulting from the inhalation; and no accident has
occurred from its use by a large number of patients
during an extended period. The daily inhalation of
twenty to forty litres, continued for the space of a
month or six weeks, gave rise to a moderate feeling of
warmth in the fauces and chest, occasionally accompanied by a slight headache. As a rule, the pulse at
first increases in the number of beats, but in some
persons the pulsations become less frequent; and the
inhalation is generally followed by increased appetite
and strength—the improvement of the former being
frequently very remarkable, while the assimilatory
powers become more vigorous. These changes are
not, however, so well marked in patients who have
been previously worn out by chronic illness. A sin-
gular alteration is observed in wounds, whether recent
or old, after patients have inhaled the oxygen for several
successive days; these become red and turgid, and sup-
purate much more freely than they had done before the
inhalation was commenced. This peculiar action of
oxygen explains why its inhalation is attended with
such unsatisfactory results in the last stage of tubercular phthisis. Patients in this condition derive considerable benefit at the commencement of the practice
of inhalation, but the inflammatory symptoms soon
become more intense (as in the case of external suppurating surfaces), and these are followed by abundant
expectoration and more urgent cough, and death would

soon occur if the inhalation were persevered in. In this way, doubtless, disastrous consequences have resulted where persons, ignorant of the physiological properties of oxygen, have directed its indiscriminate inhalation, and discredit has been thereby thrown upon oxygen as a remedial agent; but this circumstance, instead of tending to depreciate the real value of this remedy, simply furnishes a proof of the necessity which exists for proper professional advice before the adoption of inhalation, as, indeed, before following any other method of medical treatment.

The therapeutical applications of oxygen are numerous, and it may be safely resorted to, excepting when certain contra-indications, which will be enu-, merated further on, are present. Oxygen is particularly serviceable in the treatment of cases of disease attended by an anæmic or chlorotic condition, in cases of debility, and in affections which exercise a depressing effect upon the system, such as, for instance, diphtheritis, diabetes, and the secondary and tertiary forms of syphilis; in affections of this nature, if the age and general state of the patient are favorable, the inhalation of oxygen is soon followed by an improvement in strength and spirits, and often greatly increased appetite. The lips and surface of the body assume a more healthy color, greater vitality is manifest, and much of the nervous irritability previously present disappears. During the course of inhalation we must, however, especially inquire into the condition of the internal organs, and, in fact, of the whole body, because, as has already been stated, suppurating surfaces, under the stimulating effects of the oxygen, become so greatly modified in their character, that inflammatory action is eventually set up, unless the case be carefully watched. At the same time, the mere existence of a sore or a wound is not, of itself, sufficient to contra indicate the employment of oxygen : indeed, on the other hand, this remedy may be not unfrequently used with great advantage in the treatment of certain kinds of sores or wounds, which are characterized by the absence of sufficient vitality, and which consequently remain stationary, or heal only very slowly. The action of oxygen is much sooner manifested in the young than in old persons.

In the foregoing description of the physiological and therapuetical effects of oxygen, I have purposely followed the reports presented to the Academy of Sciences of Paris by MM. Demarquay and Leconte, because their researches have been more extensive than those of any other experimenters. Other observers, however, both in this country and on the continent, have arrived at similar conclusions, and the interesting discussion which followed a paper by Dr. Richardson, read at a recent meeting of the medical society of London, as reported in the February number of the "Medical Mirror," shows that the subject of inhalation is one which at the present time attracts a large amount of professional attention.

The efficiency of oxygen is undoubted in most cases of disease attended by debility, and diminution of the red corpuscles of the blood. In asthma it sometimes acts like a charm in removing the difficulty of breathing, and restoring the patient to a healthy state. In the later stages of consumption it is inadmissible, unless resorted to only occasionally and for a short period, for the reason which has been already given, but when inhaled earlier in the course of the affection, it may be productive of considerable benefit.

The inhalation of oxygen is contra-indicated in affections accompanied by much fever; in deep-seated inflammatory disorders and visceral diseases; in most affections of the heart and large vessels;* in neuralgia occurring in plethoric individuals; and when the hæmorrhagic diathesis is present.

The difficulties which formerly existed in connexion with the practice of inhalation, such as those of obtaining gases pure, and also at any time or in any place, are now almost obviated. With respect to oxygen, a portable and very useful apparatus in which it can be generated, and from which it can be inhaled, was lately exhibited before the Medical Society of London. It can be procured of Messrs. Garden and Robbins, of Oxford Street, by whom is also sold an oxygennesis-powder, with which the gas can be easily and promptly prepared.

Besides oxygen, numerous gases have been employed therapeutically, and some of them possess valuable properties.

The vapor of iodine, one of the best disinfectants known, may be occasionally inhaled with advantage by persons suffering from scrofulous affections, enlargement of the tonsils, sore throat, ozæna, and similar disorders. The readiest mode of inhaling it is from a box with a perforated lid; or a small quantity of the tincture of iodine may be added to some hot water, so that the patient can inhale the iodized steam. Dr. Murray, in his "Treatise on the Influence of Heat and Moisture," and some other authors state that

---

* In Cyanosis, the inhalation of oxygen, properly employed, is productive of great benefit. The patient improves in general health, the color of the face and lips becomes more natural, and many of the signs of imperfect oxygenation disappear under this mode of treatment.

2

they have employed iodine in the gaseous form, with much benefit to the patients. Dr. Murray says, that he has always observed improvement of a temporary kind, at least, in the condition of the patient; the cough becomes diminished in violence and the expectoration easier, while the patient sleeps better. Other observers have not arrived at such favorable conclusions as this writer, but there appears little doubt that the inhalation of the vapor of iodine forms an efficient method of administration in cases where iodine or its preparations are indicated, such as hypertrophied tonsils, etc. One of the latest observers on this point, M. Simon, states, in the "Union Medicale," 1861, that out of twenty-eight phthisical patients, seventeen derived positive benefit from iodine inhalations, both as to general and local symptoms, and four might be considered as cured. Iodine has been employed in this form also in cancer, but without any decidedly good results.

Another gas which has been highly spoken of by some writers in the treatment of phthisis, is chlorine gas, which has ever been credited by some with the cure of this affection. Louis, in his "Researches on Phthisis," states that he studied the action of chlorine on upwards of fifty consumptive patients, without any sufficiently successful results to warrant him in speaking so favorably of it as some other authors had done. It is probable that the discrepancy on this point, as upon most of those medical questions about which there is a diversity of opinion, is principally due to the fact that the cases experimented on by various observers differed either in degree or even in kind. Louis himself appears to admit that the inhalation of chlorine gas is efficacious in chronic pulmonary catarrh, (a point on which numerous authorities are agreed), although not in tubercular phthisis. It may be employed whenever direct stimulation of the mucous membrane of the air-passages is indicated.

The other gases which have been most frequently used for inhalation are nitrous oxide, carburetted hydrogen, and carbonic acid. The first named has been greatly extolled in the treatment of asthma. Carburetted hydrogen was employed by Davy, who thought that it was possible that by various combinations of nitrous oxide, "we should be in possession of a regular series of exciting and depressing powers, applicable to every deviation of the constitution from health" (vide "Davy's Chemical and Philosophical Researches"). Carbonic acid, administered by inhalation, when diluted with atmospheric air, has been recommended by various writers. According to Vogler, (*Deutsche Klinik*, 1859), the respiration of the carbonic acid, diluted with air, given off from the springs at Ems, is attended with more or less irritation, so that it is likely to be beneficial only in cases where stimulating treatment is indicated.

Air, either moist or dry, medicated with various remedial agents, has been employed therapeutically from the earliest periods of medical science, and the mere enumeration of the different substances which have been thus used would constitute a formidable looking list. The action of these medicated vapors is, of course, almost analogous to the effects produced by the same remedies when administered by the mouth, and they are consequently indicated in a similar manner, according to the individual requirements of the case, and to the special indications for stimulants, anti-spasmodics, or sedatives. I must, therefore,

content myself with a brief mention of the principal
remedies which are, or have been, administered by
inhalation.

Arsenical fumigations were held in high esteem by
the ancients in the treatment of asthma, bronchitis,
catarrh, and some other affections of the air passages
and lungs. The preparation of this mineral which
they employed was a more inert substance than the
arsenic employed in modern times. It was almost
identical in composition with yellow orpiment, and is
supposed to have consisted of sixty-two parts of
arsenic and thirty-eight of sulphur, according to Kla-
proth (vide "The Sydenham Society's Translation of
the Works of Paulus Ægineta," vol. i., p. 480). Arsenic is seldom employed in inhalation at the present day,
but cigarrettes which have been steeped in a solution
of arsenious acid are occasionally recommended to be
smoked by asthmatic persons. Dr. Leared has published some cases in which signal benefit was obtained
from smoking these cigarettes.

The ethereal preparations, especially sulphuric ether,
are sometimes useful as sedatives and antispasmodics,
when inhaled. A few whiffs of air, containing a small
proportion of chloroform, will be found very serviceable
in allaying or removing the irritable cough of spasmodic asthma.

The inhalation of the vapor of hot water containing
camphor, conium, belladonna, hyoscyamus, lobelia,
stramonium, and other vegetable sedatives, is attended
with very beneficial results in all diseases of the chest,
in which there is much local irritability and troublesome
cough.

Creosote and tar have been recommended for use

in a vaporized form, when stimulating remedies are indicated.

Nitrate of potash fumes have long been employed in a similar class of cases; and those of hydrochlorate of ammonia have been more recently well spoken of in cases where a stimulating plan of treatment is necessary. In using nitrate of potash, pieces of blotting-paper, previously soaked in a saturated solution of the salt, and then dried, are burnt upon an earthenware plate; the fumes are soon diffused throughout the room, and their beneficial effects are, in many cases rendered evident, as Dr. Watson observes, when writing of the value of this remedy in asthma, "in clearing out the passages, and gradually opening the air-tubes." The inhalation of the fumes of hydrochlorate of ammonia is recommended in pulmonary catarrh by Paasch; they may be easily generated by pouring a little hydrochloric acid into a watch-glass placed in a saucer containing some liquor ammoniæ.

The different balsams and gum-resins which possess expectorant properties may be volatized by heat, and their vapors inhaled, with good results, in asthma and bronchitis. Those most commonly employed in this manner are, the balsam of tolu, balsam of Peru, benzoin, and storax.* Their efficacy may be increased by the addition of a little spirit.

* As a direct proof of the antiquity of the practice of inhalation, the following passage may be quoted from the writings of Paulus Ægineta who lived at an early period of the Christian era :—" To be inhaled for a continued cough : storax, pepper, mastick, Macedonian parsley, of each one ounce ; sandarach (an arsenial preparation), six scruples ; two bay-berries ; mix with honey ; and fumigate by throwing them upon coals so that the person affected with the cough may inhale the vapor through a funnel. It answers also with those affected by cold in anywise." The chief objection to this prescription is, that there are too many ingredients contained in it.

The benefit obtained by the inhalation of moist medicated vapors is attributable, in some measure, to the
soothing effects of the steam of the hot water in which
the various remedies are dissolved. I am in the habit
of recommending patients suffering from congestive
asthma, bronchitis, and analogous pulmonary affections,
to inhale at intervals the simple vapor of heated water,
at different degrees of temperature, according to the
nature of the case, without the addition of any medici-
nal agent. The warm, moist vapor is most agreeable
to the patient, and seldom fails to afford immediate
relief. The expectoration becomes freer, the cough is
less frequent, and the dryness and irritation of the air-
passages are entirely removed. The inhalation of hot
steam is very efficacious in the treatment of hay-asthma or hay-fever, as I have already pointed out, in a
communication to one of the medical periodicals, on
the subject of this troublesome affection.

There are numerous methods by which the patient
may be enabled to inhale the steam, as for instance,
from a basin or jug containing hot water, or by breathing through a sponge dipped in hot water, and par-
tially wrung out. Several forms of apparatus have
also been devised for this purpose. The most efficient
apparatus for the inhalation either of simple steam or
of medicated vapors is that which is known by the
name of Nelson's Inhaler; it is constructed of earthenware, and, in addition to its complete adaptation to
the purpose for which it is intended, possesses the
triple recommendation of cleanliness, portability, and
cheapness.*

Until a very recent period, only such remedial sub-

* This useful apparatus is manufactured by Messrs. Maw & Son, of
Aldersgate Street.

stances could be used in inhalation as could be administered in the gaseous form, or volatilized into vapor, so as
to allow of their being drawn into the lungs during the
act of inspiration. During the last few years, however,
the discovery of the laryngoscope has led to the closer
investigation of the diseases of the larynx and air-passages generally, and various new methods of applying
local medication have been devised. Of these the most
ingenious, and, at the same time, the most important,
are those which have for their object the minute subdivision of different remedial substances, so that they
can be inhaled, when in solution in the form of spray.
The first apparatus invented for this purpose was that
of Dr. Sales-Girons, but it has been since modified and
improved upon by several individuals, particularly
Lewin, of Berlin, and Siegle, of Strasbourg. The
instrument invented by Sales-Girons in 1858, is so constructed that the medicated fluid is forced, by the
agency of compressed air, through a tube having a
very small opening against a metal plate. At this
point the stream of fluid is checked, and it becomes
divided into fine spray, (to which the term of pulverized, or atomized, has been applied), and in this condition can be inhaled by the patient. A modification of
this has been introduced by Bergsen, of Berlin, whose
apparatus consists of two glass tubes, having capillary
openings at one end—these two ends being placed
almost at a right angle with each other. The more
open end of the perpendicular tube is immersed in the
medicated fluid, and, as the compressed air is forced
through the horizontal tube, the air in the perpendicular one becomes exhausted, and the medicated solution
then rises in it, and, when it arrives at the capillary
opening, is dispersed in very fine spray by the force of

the compressed air passing along the other tube. The
principle upon which this instrument acts is familiarly
illustrated by the perfume-odorators which have been
introduced into general use during the last few months,
and which are, in fact, merely an adaptation of Berg-
son's ingenious invention to common purposes. Siegle's atomizer is constructed on a similar principle, with
the substitution of steam for compressed air as the
means of dispersing the medicated liquid. It has been
doubted whether solutions, thus divided into minute
spray, reach the lungs, but the fact that they do so has
been incontestably proved by numerous observers. In
one striking case, experimented upon by M. Demarquay,
it was shown that the inhaled fluid reached the trachea.
The case was that of a woman who had a tracheal fistula, and in whom, after she had inhaled the spray from
the pulverizer, the introduction of chemical tests
through the tracheal opening proved the presence of
the various substances which had been inhaled, thus
clearly proving that, at any rate, the spray which had
been drawn during the act of inspiration had reached
as far as the fistulous opening. Numerous other proofs
of the entrance of the medicated spray into the air passages, and even the lungs themselves, have been brought
forward. The most satisfactory proofs of this fact are
the practical results which have been obtained by the
administration, in this manner, of different remedial
agents, and the marked benefit which has resulted from
their employment in hæmoptysis, asthma, bronchitis,
whooping-cough, and some other pulmonary affections.
The principal medicines which have been used in this
way, are nitrate of silver, alum, tannic acid, iodide and
bromide of potassium, bicarbonate of potash, tincture
of the sesquichloride of iron, sulphate of zinc, acetate

and hydrochlorate of morphia, tincture of opium, the extracts of hyoscyamus and conium.*

With respect to the inhalation of dry air, as distinguished from that of moist vapors, two plans of treatment demand notice, viz., that by compressed air, and that of the inhaling tube, invented by Dr. Ramadge many years since.

Of the former I have no personal knowledge. The idea of employing this novel mode of treatment is stated in the "Gazette Medicale de Paris," to have originated in the circumstance that several of the work-. men who were engaged in sinking the foundations of a bridge, and who happened to be affected with various chest diseases, were comparatively free from the symptoms of their complaints when they were at work in the caissons sunk below the surface of the water. Establishments at which patients are subjected to the effects of condensed air have been formed at Paris, Lyons, Montpelier, and some other continental cities; and the promoters of this plan of treatment describe it as more or less successful in cases of catarrh, chronic bronchitis, asthma, anæmia, etc. The pressure used averages from $1\frac{1}{4}$ to $1\frac{3}{5}$ atmospheres. I must own that I am very skeptical as to the benefit alleged to be derived from treatment by compressed air, particularly as numerous cases have been put on record both in this country and abroad, of the production of paralysis, and other serious consequences, in laborers who have

---

* Since the above given paragraph was written, the author has seen a report (Lancet and Medical Times, Feb. 25th) of the meeting of the Medico-Chirurgical Society, at which a paper by Dr. Mackenzie "On the Inhalation of Atomized Liquids" was read. Dr. Mackenzie and Dr. Gibb, both of whom are able observers, were agreed as to the value of inhalation in the treatment of certain affections of the air passages and lungs.

3

remained subject for too long a period to the effects of compressed air. Those who are desirous of complete information concerning this plan of treatment will find a long paper upon it, by Dr. Sandahl, in Schmidt's "Jahrbuch" for 1863.

The inhaling-tube, after having crossed the Atlantic, has been re-introduced into this country, and made widely known through the medium of monster announcements in the daily journals, by an individual whose sole and very slender claim to any knowledge of the treatment of pulmonary affections is based on the mere circumstance of his having brought out an English reprint of a book written by an American author. This tube is constructed in such a manner that air can be easily drawn through it into the lungs, but is not so quickly expired. The normal balance between inspiration and expiration is thus re-established, and in suitable cases for this kind of inhalation, the chest becomes gradually expanded, so as to resume more natural dimensions. Dr. Ramadge says that is a valuable adjunct to other treatment in consumption and asthma.

II.

THE APPARATUS FOR INHALATION.

A vague idea of the benefit of inhalation in the treatment of diseases of the air passages is present with nearly all physicians; but it rarely assumes any tangible form, because of the dearth of information in the text books. Occasionally we have a spasmodic manifestation of it in the use of the vapor of water or decoction of hops, tansey, etc., in acute disease, inhaling the fumes produced by burning nitrate of potash in asthma, or of resin, Canada balsam, balsam of Peru, tar, etc., in bronchitis and phthisis. But though occasional benefit results, the attempts are so crude they are soon given up.

The oldest method of employing inhalations, was, to vaporize the material employed on a hot iron. Thus any fluid might be employed, or any solid, like the resinous or balsamic agents, brought into a condition that they might be inspired. Another method, in the case of fluids, was, to place them in a vessel and by the aid of a hot iron, brick, or stone, to produce the necessary amount of vapor. Though the methods were crude, I have seen very marked benefit result from their use. The employment of the vapor of water would rarely fail to soothe irritation in disease of the larynx and bronchii, and when this induced contraction of the muscular tissue and consequent difficulty in respiration, would give temporary relief until it could be rendered permanent by the use of other means.

Volatile agents were used in many ways. Many years back, I recollect seeing them put in a common wine bottle partly filled with hot water, and the vapor inhaled as it passed from the mouth of the bottle. Others would employ a coarse sponge. Pressing it out of hot water, the medicine was sprinkled upon it, and breathing through it the patient obtained the vapor of the medicine, and a moist air which was not unfrequently of full as much benefit.

Many forms of apparatus are employed at present, each being thought preferable by its inventor. We need describe but four of these, as they will fulfill every indication. As the remedies employed may be divided into two varieties, *volatile* and *non-volatile*, two forms of apparatus will be indispensable.

The volatile agents are vaporized by heat, and a majority of them at a temperature less than that of boil-

Fig. 1.

ing water. It requires then a container to receive the
fluid, and a tube to conduct the vapor to the patient.
Fig. 1, represents such an instrument. The flask is
made of flint glass, and will usually bear an elevated
temperature. The cork is perforated by an opening for
the attachment of the rubber tube, and for a smaller
glass or metal tube for the admission of atmospheric
air, as the patient inhales the vapor. Nelson's Inhaler,
spoken of on page 22 is also a good instrument.

In place of this, I have used the apparatus represented
in Fig. 2. It consists of a tin cup, perforated at the

Fig. 2.

bottom and a three quarter inch tube inserted and sol-
dered. To this, is attached two or three feet of rubber
tubing, which is terminated by a mouth and nose-
piece as represented in the wood cut. The cup contains two cross wires to hold the sponge, which should
be coarse and open. In using this apparatus, the
sponge is pressed out of hot water and put in the cup,
or it may be wet in the cup, and the medicine then
sprinkled on it in sufficient quantity. The cup being
placed upon the floor, a chair, or, if more heat is
required, upon the stove, the patient inhales the vapors
as they arise. In country practice, it is well to have
the cup made the size of a tea-kettle, or tin teapot lid,

so they can be turned over these vessels if it is desirable to inhale the vapor of water. This apparatus
recommends itself, in that it is simple, cleanly, and
cheap, being easily manufactured wherever a tinner can
be found, at a cost of 75 cents.

The volatile agents, however, form but a small portion of those we should like to use, and we must, therefore, have some apparatus for conveying to the air passages the remedies themselves. Such an apparatus we
now have in great perfection, evenly distributing any
fluid, no matter what its character. They are known
by the generic name of *Atomizers*, inasmuch as they
minutely divide the fluid, that it may be inhaled.
Though there are a number of these, it will be necessary to describe but three.

The simplest instrument is that known as "Elsberg's
Nebulizer," which consists of two hard rubber tubes
pointed at the extremities, the openings being small,
and so hinged that they can be placed at right angles,
the openings being immediately opposite, as in Fig. 3.

FIG. 3.

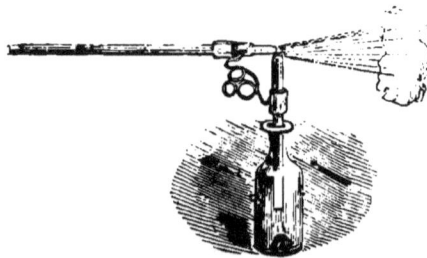

One arm of the apparatus, being placed in the medicated
fluid, blowing through the other, causes the fluid to
rise in the tube, and it is carried off in a fine spray.
Rimmel's *Rafraichisseur*, which is the same in principle, has been employed for some years for distributing
perfumes, and may be purchased quite cheap. The

principal objection to this method is, that it requires a second party, and the breath cannot but prove offensive to many patients.

The second form of apparatus, consists of a cylinder in which works an air tight piston, like the barrel of a syringe. Fluid being placed in it, is forced through minute openings in the nozzle, as a delicate spray.

FIG. 4.

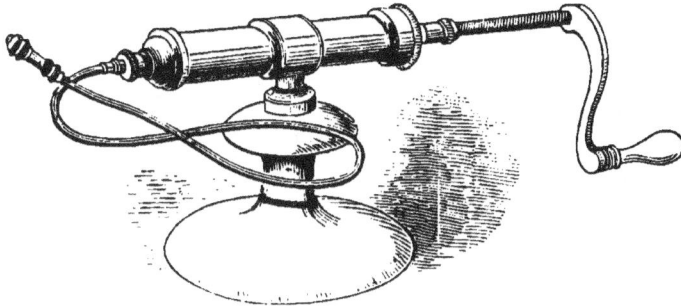

Fig. 4, represents the instrument of M. Sales Giron, which I have used in my practice with excellent results. When inhalations are much used, I have no doubt they will be manufactured by our hard rubber manufacturers, at a price to bring them within the reach of all.

The apparatus of Dr. Mackenzie is a very good one. The piston is drawn back by a wheel and rack at its upper part, and is forced down by a circular spring which surrounds the cylinder. The apparatus is filled with liquid by a funnel in its top, and all the spray, except that which is inhaled, passes back into the apparatus. He claims the following advantages for it.

" 1. Its simplicity, requiring only a few turns of a handle to set it in operation. 2. The extremely fine state of subdivision which it effects. 3. The uniform pressure exerted. 4. The fact that the waste liquid returns into the apparatus.   5. The ease with which it can be taken to pieces and cleaned."

Two or three other varieties are used, but the principle is the same.   In the French instrument, the piston is forced down with a screw, and the liquid forced out with very great pressure.   Four or five turns with the wheel is sufficient to afford a jet of spray for as many minutes.

With these instruments, the medicated liquid is minutely broken up, and the spray inhaled with the air.   In fact, the division is so fine, that it will pass wherever the air goes.   The temperature of the fluid may, to some extent, be adapted to the case; but the general impression, whether the fluid is used hot or cold, is that of coolness.

FIG. 5.

The third form of apparatus is that of Dr. Seigle, and is far preferable to the others, for its simplicity and because it is automatic.   The best reason for preferring it, however, is, that its price is such as to bring it within the means of any patient, as it is furnished through the druggists for $5,00, and its construction is so simple, that it is readily operated by any one.

The instrument consists of a little kettle, into which

is screwed a fine cork perforated with a horizontal tube,
in which there is a fine opening. Placed at right
angles to the horizontal spout, is a vertical tube, which
dips down into a small cup containing the medicated
fluid. As the steam issues from the horizontal tube, it
causes a vacuum in the vertical tube, and the medicated
liquid rising up becomes mechanically incorporated in
the steam, and is blown off in the form of a minutely
divided spray. The dilution of the medicated liquid
which takes place is very slight, as the conversion of a
drachm of water into steam will take up three drachms
of medicated liquid. The temperature of the steam is
lowered by the incorporation of the liquid, so that at
the end of the cylinder, it has only a temperature of
seventy degrees. The instrument represented in Fig. 5,
is manufactured by Mr. Max Woeher of this city.

"In this apparatus there is, of course, no current of
cold air. The amount of liquid taken up varies, that
is, it depends on the amount of heat applied, on the
height of the column of liquid, etc. This is not an
important defect; but when it is desired to take up a
definite quantity of liquid, the author uses the following apparatus: A graduated glass tube, about eight
inches high, has from its lower part a fine piece of
tubing, which is bent round and up again, and then
extends about an inch horizontally, and ends with a
minute opening. In the vertical portion of the fine
tubing, there is a stop-cock. The small apperture of
the tube is bent at right angles to the tube from the
kettle, and as the liquid emerges, it becomes incorporated in the steam. By means of the stop-cock, the
amount of liquid which passes from the tube can be
regulated, so that the same amount can always be taken
up at the same time."

3

I have employed this instrument in every case where inhalations could be used with the least prospect of advantage, and feel justified in recommending it to the profession. There can be no doubt, but that the medicated fluid is carried through all parts of the respiratory tubes. This diffusion is an objection in the treatment of some affections, as we would like to restrict the remedy to the diseased surface. Still as we learn more of the action of medicine upon the body, we will care less about its concentration, and more about selecting the appropriate remedy.

I also employ a modification of Seigle's apparatus in city practice. It consists of a small boiler, with two openings at the top; one for the insertion of the glass tube in a cork, the other closed by a cork, for filling the boiler with water. This boiler is placed on a Musgrave's, or other small gas heater, which furnishes the heat. It answers an excellent purpose, and is cheaper.

FIG. 6.

Another apparatus of like character is constructed upon the plan of the Kerosene Lamp Heating Company's boiler. Coal oil is employed, and in these days of high taxes and dear alcohol, the use of coal oil in these instruments is an object. I think it decidedly the best atomizer in use, not only for its cheapness, but for its safety, and the fact that we can get a much stronger jet of steam from it than from the original Seigle apparatus: this is sometimes very desirable. Fig. 6 represents the apparatus, which may be obtained of all prominent druggists, or by addressing J. G. Henshall and Co., Cincinnati. The price is $5,00, securely packed for shipment.

Dr. Mackenzie, who has experimented to a considerable extent with atomized liquids, thus sums up the results, in a paper to the Medical Times and Gazette:

"Leaving Demarquay's rabbits out of the question—it having been shown by Claude Bernard that as those animals in their normal state breathe through the mouth, the conditions are not physiological; and by Fournie that any solution (not atomized) injected into a rabbit's mouth passes into the lungs—there are: 1. Demarquay's and Brian's experiments on dogs. 2. His (Dr. Mackenzie's) on pigs and dogs. 3. An experiment performed by Demarquay, in the presence of numerous witnesses, on a woman with a tracheal fistula, in which it was shown that the inhaled liquid penetrated to the trachea, though there was a great obstruction at the upper opening of the larynx. This experiment, which had been previously unsuccessfully performed by Fournie, has since been repeated by Lieber, Schnitzler, and others, with results similar to those obtained by Demarquay. 4. The fact first shown by Bataille, and since by Moura-Bourouillon, the author, and others, that after the inhalation of a colored atomized solution, the sputa remained tinged long after the employment of the laryngoscope could detect any traces of the material used.

"On the one hand, there were an immense number of positive proofs of the penetration of atomized liquids, on the other hand there were a few experiments performed with negative results. It was scarcely necessary to remark that any experiment might be performed—the most simple chemical test employed—in a manner to insure failure. But a few experiments of this sort could have little weight against the mass of evidence on the other side.

"The author stated that the greatest benefit from this system of therapeutics might be expected, and had resulted in bronchitis, asthma and hæmoptysis. He brought forward twenty-two cases treated between October, 1863, and January 1864. There were ten cases of bronchitis, six of phthisis, two of hæmoptysis, three of asthma, and one of hooping-cough. The author did not believe that in phthisis the treatment would have a positively curative effect, but was beneficial in cutting short intercurrent bronchitis. Of the twenty-two cases detailed, only two were unable to make use of this curative process. Of the ten cases of bronchitis, eight were cured, one relieved, and one obtained no benefit.

"The average duration of the time required for curing these cases, though most of them were severe and of long standing, was only fifteen days and a quarter. The shortest time was six days (a severe case, No. 4); the longest forty days. The duration of treatment was not in proportion to the severity of the disease, one mild case requiring twenty-eight days to get well.

"Of the six patients laboring under consumption, two were unable to use the inhalations on account of the

irritation which they caused. . Of the remaining four
cases, whilst the physical signs did not undergo any
material alteration, the local symptoms (expectoration,
pain, and cough) were greatly relieved. The general
health was much improved in two cases. Nos. 11 and
15; slightly in a third, and not at all in a fourth. In
two cases of hæmoptysis, one severe, the other slight,
the atomized liquids rapidly stopped the bleeding. In
three cases of asthma—one very severe case, which had
obstinately resisted the ordinary treatment—this system
of therapeutics soon gave relief. In one case of hooping cough (in an adult) the inhalations gave immediate relief. In one case of hooping cough (in an adult)
the inhalations gave immediate relief, and quickly
effected a cure. The author stated that during the past
year he had used atomized liquids in more than eighty
cases of diseases of the lungs, and that he had found
the plan of treatment no less successful than was detailed in this paper. The various instruments referred
to in the communication were brought before the society, and likewise diagrams illustrating their action
and method of employment."

III.

ON THE THERAPEUTICS OF INHALA- TIONS, WITH FORMULA.

It seems almost like a work of supererogation to describe the topical action of remedies upon mucous membranes, and yet the subject of topical medication is so new, that many physicians have not thought upon the subject sufficiently to apply the general principles of therapeutics with which they are familiar.

The mucous membrane of the air passages does not differ materially in its structure, or its vascular and nervous supply, nor yet in its function, from other parts of the general mucous lining of the body. Hence, what we know by experiment and experience with regard to the topical action of remedies upon mucous structures, is applicable here. And as the principal difference between the skin and mucous membrane is in its epithelium, it will be found that remedies act in a very similar manner upon both. Thus, each one may start with a very fair knowledge of the action of remedies, and he will rarely find himself mistaken.

Of course, it is of much importance to determine the exact condition of the part affected. As regards its circulation, dryness or increased secretion, tension or relaxation, etc., giving the same care in the examination as we would to diseased surfaces that we can see and wish to medicate. Auscultation gives us this information very accurately, and in a short time the careful observer will have no difficulty in determining these points. The general symptoms are also an important guide, as the mucous membrane does not differ materially in its general condition from other parts of the system. Thus we would never expect to find a tense, dry mucous membrane, with a relaxed and flabby skin, nor a dry, irritable condition of the air passages with a soft and feeble pulse.

Whilst the general and local disease resembles each other to such an extent that the one may serve to some extent to define the other, they should be disassociated in the treatment. The general treatment will have reference to digestion, the state of the blood, excretion and innervation. It generally resolves itself into the use of the most feasible means for improving nutrition of textures, and the more active the nutritive processes become, the better are the chances for complete recovery. Thus a phthisis is many times successfully treated by the use of bitter tonics, iron, and a nutritious diet, care being taken to keep the secretions free. Added to this, the restoratives, preparations of phosphorus, cod-liver oil, etc., with out-door exercise and pleasant society, and the general treatment is as near perfect as we can expect to render it with the means at our command.

The local treatment is directed more especially to the relief of irritation, change of secretion, and to restore tone and strength to the mucous membrane. The remedies used, might be appropriately classified as *relaxant or emollient, narcotic, stimulant, tonic,* and *astrin-*

gent, though we will find it somewhat difficult to properly arrange all of them under these heads.

Relaxant or *emollient* inhalations are employed when there is irritation with dryness, or a tenacious mucus that is raised with difficulty; a condition similar to that for which we would apply a poultice externally. The vapour of water is the type of this variety, and will be found useful in the treatment of all acute inflammations in the first stage. A decoction of hops, of poppy heads, of German chamomile, or tansy, is sometimes better than water alone. If it is desirable to produce great relaxation, I usually employ an infusion of the herb lobelia, or a small portion of sulphuric ether. If there is a tenacious secretion difficult of removal, the addition of common vinegar proves valuable: in these cases, I usually order vinegar and water equal parts.

These inhalations may be employed by any mode that the fluid can be vaporized, and even the crude methods first described give excellent results. Of course the apparatus of Seigle will be better than these.

By a *narcotic* inhalation I mean one that relieves irritation of the nerves distributed to the mucous membrane, and thus checks cough. The name is not a good one, as these remedies rarely produce that deadening of sensibility that might properly be called narcosis, yet as it is the only term by which the action above named is known, I use it for want of a better. The term *sedative* has been employed, and might still be used were it not now so generally applied to those remedies that control the circulation.

By a narcotic inhalation we understand, then, a remedy which, when locally applied, relieves irritation of the part and checks cough. They are sometimes used for this purpose alone, but are more frequently added to other inhalations to effect a double purpose. Thus, added to water, or simple infusions, they are emollient-narcotic; to tonic preparations, tonic-narcotic, etc. ; and as the relief of irritation and cough is one of the prominent objects of treatment they are extensively employed in this way. It must not be supposed, however, that they are the only remedies for the relief of a cough, for many times we obtain more speedy and better results in this respect from the use of stimulants, tonics and astringents.

This class of remedies divide themselves into the volatile and non-volatile, the first being employed with the apparatus described in Figs. 1 and 2, the second class either by vaporization or by the apparatus for atomizing fluids. Almost every narcotic or sedative may be employed in this way, and upon exactly the same principles that would govern their local application elsewhere. The following will be found good formulæ for the volatile agents :

No 1.—℞ Hydrocyanic acid dilute, f℥ij; wine of ipecac, paregoric, aa f℥ss ; tincture of conium, f℥ij ; rose water, f℥xij. Half an ounce of this may be inhaled three or four times a day.

No. 3.—℞ Cyanuret of potash, gr. viij; tincture of ipecac, tincture of lobelia, aa f℥vj ; tincture of stramonium, f℥j ; rose water, ℥iv ; M. Inhale a teaspoonful every four hours.

No. 4.—℞ Acidi hydrocyanici diluti, min. xx ; tincturæ hyoscyami, tincturæ lupuli, aa f℥j ; aquæ calidæ, ad. f℥viij : M. (Tanner.) Employed in phthisis, ulceration of the larynx, etc.

No. 5.—℞ Acidi hydrocyanici diluti, min. xv ; spiritus chloroformi, f℥iij, aqua bullientis, f℥viij : Mix. In laryngitis, œdema of the glottis, etc. (Tanner.)

No. 6.—℞ Tincturæ chamomilla, f℥ss ; morphia sul. gr. x; æther, alcohol, aa f℥j : M. Use two drachms for an inhalation.

No. 7.—℞ Tinct. conii, tinct. strammonii, aa ℥ss; spiritus chloroformi ℥j : Use two drachms for an inhalation.

These formula might be increased, but sufficient has been given to illustrate the mode of combination in most frequent use. When employing fluids with an atomizer we need not be so particular as to their form. Tinctures added to water or proof spirit give the most eligible preparations, unless we except the aqueous extracts rubbed down with water. We do not combine opium with the other narcotics, as it is antagonistic in its action to most of them. Indeed, it will be found best to use each agent separately. The following will be found good preparations.

No. 8.—℞. McMunn's elixir of opium, f℥ij., decoction of hops, ℥iv. Mix. Of this half an ounce may be employed as an inhalation. It acts as a stimulant to the respiratory organs, and hence is never employed when there is dryness, but answers an excellent purpose, when there is relaxation of the bronchial mucous membrane with increased secretion.

No. 9.—℞. Acetum opii f℥ij., acetum lobeliæ, f℥j., aquæ rosæ, ℥ij. Mix. Let half an ounce be inhaled every three or four hours. This is my favorite formula for the use of opium.

No. 10.—℞. Extractum conii, ℥j., spiritus chloroformi, ℥j., spiritus vini rectificati, ℥ij. M. Add two drachms to two of water, and let it be inhaled.

No. 11.—℞. Tinct. stramonii, ℥j., aqua rosæ, ℥v. Mix. Use half an ounce for an inhalation.

Stimulant inhalations are employed in chronic laryn-

gitis, bronchitis, and phthisis, where there is relaxation
of the mucous membrane and increased secretion. They
not unfrequently relieve the cough and unpleasant
sensations, when the narcotics would have no influence.
The simpler the agents, the better their influence as a
general rule.

No. 12—℞. Vinegar of lobelia, f℥j., comp. tinct. of
lavender, f℥ij., infusion of chamomile, f℥v. M. Use
half an ounce with the atomizer every four hours.

No. 13.— ℞. Tinct. iodine, f℥ss., Rectified spirit,
f℥iij. Mix. To be used in the same way. Care should
be employed at first not to get the inhalation so strong
as to produce irritation, and the vapors breathed
through a glass funnel prevent soiling the face and
clothes. If found beneficial the strength may be gradually increased.

No. 14.—℞. Tincturæ iodi., min. xxx., aqua calidæ,
f℥iv. Mix. To be cautiously inhaled with a common
nhaler. (Tanner.)

No. 15.—℞. Olei terebinthinæ, f℥j., aquæ calidæ,
℥vj. Mix. To be employed in the same way in
:ases of chronic bronchitis with excessive secretion.
Tanner.)

No. 16.—℞. Creosote, min. xxx, aquæ, ℥iv., tinct.
avendulæ comp., f℥ij. Mix. May be employed in
either way.

No. 17. ℞. Nitrate of silver, gr. xx., distilled water,
℥ij. Mix. This should be used with the glass funnel,
and its strength carefully proportioned to the susceptibility of the patient.

No. 18.—℞. Permanganate of potash, gr. xv., distilled water, ℥iv. It may be best to commence with a
weaker solution than this, but in some cases the strength
may be increased to ten grains to the ounce of water.

It will be found very serviceable in cases where there is profuse expectoration.

No. 19.—℞. Solution of chlorinated soda, f℥ij., rose water, f℥iv. This also will have to be employed with care.

No. 20.—Lime water of full strength, with or without the addition of morphine, conium, or other narcotic, is an excellent remedy in cases, where expectoration is difficult.

The bitter tonics are employed when there is debility and relaxation, and very frequently serve as vehicles for the use of the narcotics. The infusions seem to answer a better purpose than the tinctures, but either may be used. I have employed the German chamomile with much advantage, and also the hydrastis and the chiretta. An infusion of Peruvian bark answers an excellent purpose in many cases, and I not unfrequently order the tincture. An infusion of senega and serpentaria may also be used. The solution of persulphate of iron and perchloride of iron is occasionally useful, but gener ally proves too irritating.

No. 21.—℞. Quiniæ sul. gr. xx., dilute sulphuric acid, f℥ij., acetum opii, f℥ij., infusion of hops, ℥iv. Mix

Astringents are not very frequently employed, but may be occasionally used with advantage. A solution of tannic acid will be found as useful as most of the vegetable astringents.

No 23.—℞. Tannic acid, ℨj., quinine, gr. x., aromatic sulphuric acid, ℨij., rose water, ℥iv. Mix. Use half an ounce for an inhalation.

No. 24.—℞. Alum, ℨij., tincture of cinnamon, ℨij., water, ℥iv. Mix. To be used as the preceding.

No. 25.—℞. Red oak bark, yellow dock root, aa. ℥j.. Make an infusion with two pints of boiling water and use as above.

The following are the inhalations employed at the Consumption Hospital London, as given in Squires' Pharmacopœia's of the London Hospitals. They belong to the class of volatile inhalations, and are employed with the apparatus represented in Figures 1 and 2.

No. 26.—Inhalatio Acid Hydrocyan. Dilute Hydrocyanic Acid 10 to 15 mins.; for one inhalation.

No. 27.—Inhalatio Æther. Chlor. c. Hyoscyam. Chloric Æther 30 mins.; Tincture Henbane 30 mins.; Infusion of Hop, or water 8 oz.

No. 28.—Inhalatio Camphoræ. Spirits of Camphor 1 to 2 drms.; boiling water 8 oz.

No. 29.—Inhalatio Chlorinii. Chlorinated Lime 2 oz.; for one inhalation.

No. 30.—Inhalatio Chloroformi. Chloroform 15 mins.; for one inhalation.

No. 31.—Inhalatio Coniæ. Coniine 1 gr.; Spirits of Wine 10 mins.; water ½ oz.

No. 32.—Inhalatio Creasotum. Creasote 6 mins., water ½ oz.

No. 33.—Inhalatio Iodinii. Iodine 2 grs.; Spirits of Wine 10 min.; water ½ oz.

No. 34.—Inhalatio Lupuli. Hops 1½ oz.; boiling water 20 oz.

No. 35.—Inhalatio Opii. Extract of Opium 3 grs.

In using inhalations we are guided almost entirely by the sensations of the patient, as regards their frequency, duration, and continuance. If it gives relief from the cough, and the patient breathes freer and easier, it is doing good, but if the breathing becomes more difficult, though the cough may be relieved, it is doing harm.

IV.

INHALATIONS IN SPECIAL DISEASES.

We employ inhalations in affections of all parts of the respiratory mucous membrane, and contrary to the general opinion, they are found more markedly curative in some acute than in chronic diseases. As heretofore remarked, they are principally palliative in chronic disease, and must be associated with appropriate general treatment. It will not be necessary to consider each affection at length, but only as it is influenced by inhalations. These will replace those remedies that are employed for their topical influence, but will in no other respect interfere with the treatment commonly pursued. What is said, therefore, may be regarded in the light of addenda to our works on practice.

With a strong desire to present the subject in such a light that it will be found to stand the test of experiment, it is not improbable that it may occasionally be too highly colored. This cannot be avoided, when a subject is new and experience comparatively limited.

CATARRH.

In confirmed catarrh, inhalations have proven of marked advantage, and in some cases curative. The present winter we have had an extended experience in this disease, and whilst in some cases they seemed of little benefit, in others they surpassed the usual means. Of course, in the early stage of the affection, I should order instead, two grains of Opium at bed-time, or fifteen grains of Bi-carbonate of Ammonia, or a full dose of Tinct. Gelseminum, with rest until noon of the next day. These are means that are speedily curative, and quite certain in their action. Yet when the case has progressed until the mucous membrane is thickened, with abundant secretion of mucus or muco-pus, we do not expect this speedy influence.

In this case I frequently order an inhalation of lime water, every three hours with Seigle's apparatus, especially if the secretion is tenacious and removed with difficulty. An inhalation of vinegar and water, or of an infusion of chamomile, or hops, does very well. Formula No. 9 is an excellent inhalation when there is much fullness with closing of the nose. Formula No. 16 will be found useful in protracted cases when the discharge is abundant.

OZÆNA.

In the treatment of ozæna, inhalations will be found to be but palliative in the most of cases, yet they have proven of sufficient importance to stimulate further investigation. The most efficient means of bringing remedies in contact with the diseased mucous surface, is by the new hydrostatic method, or by the use of the common pump syringe. Any remedy may be thus directly applied, and in such quantity as is necessary.

The use of inhalations of water, water and vinegar, or some of the simple infusions spoken of in the first part of the preceding chapter, answer an excellent purpose in allaying irritation. Lime water, solution of Chlorate of Potash, or solution of Chlorinated Soda diluted, arrest the fetor of the discharges, and act as gentle stimulants. The tonic and astringent inhalations will occasionally be found useful.

CHRONIC PHARYNGITIS.

In chronic pharyngitis, especially if the disease extends above the soft palate, inhalations will be found useful. Occasionally these cases are found very stubborn, and cause much annoyance. The local treatment with nitrate of silver is frequently effectual, but in many cases, the affected parts are not reached, and the disease continues. The use of the universal syringe with the curved perforated tube furnishes a good means to make local applications behind and above the palate, and may be employed with advantage.

In using inhalations, I prefer, if the patient can, that they be drawn into the mouth, and forced out through the nose. If they cannot be used in this way, they should be inhaled partly by the nose and partly by the mouth. The tonic infusions with Chlorate of Potash are the agents most frequently beneficial. Formula No. 23 and 24 will be useful when the muocus membrane is relaxed and flabby.

TONSILLITIS.

In the treatment of acute inflammation of the tonsils. I have found inhalations, with Seigle's apparatus, more speedily beneficial than any other method of treatment. Every physician in active practice knows the almost entire uselessness of the common means employed, and to the sufferer from frequent attacks, any means that would offer a prospect of relief would be gladly welcomed.

I generally order an inhalation of a strong infusion of German Chamomile, or of one part of the tincture to six of water. Tincture of Aconite, (the root), thirty drops to half an ounce of water is also very good. A decoction of hops or tansy may be employed instead of these, or in some cases, Formula 9 or 10. The inhalation should be continued from five to ten minutes, and repeated sufficiently often to give the patient ease, if the case is severe. In milder cases two or three times a day will be found sufficient. If there is much febrile action I order small doses of Aconite, as—℞ Tincture Aconite, rad. ʒj; Aqua, ℥iv: M. Give a teaspoonful every one or two hours.

CYNANCHE MALIGNA.

In this affection, inhalations will be adjuvant to the general treatment, as is all local means. The relief of the stomach by a thorough emetic and the use of quinine and iron, with chlorate of potash or sulphite of soda are the curative agencies employed. Still as the patient suffers greatly from the throat, and has diffi-

4

culty in using a gargle effectually, we will find inhalations of sufficient importance to employ them.

We have here an atonic condition of the mucous
membrane, a feeble circulation and innervation, and arrest of the nutritive processes. Our topical applications must therefore be stimulant, and as far as possible of such character as will arrest the destructive met-
amorphosis so rapidly going on. An inhalation of a
saturated solution of Chlorate of Potash, or Sulphite of
Soda answers an excellent purpose. I have found more
benefit, however, from a strong infusion of Baptisia
Tinctoria in equal parts of vinegar and water, than from
any other means. This is also an excellent gargle.

DIPHTHERIA.

In this affection, the employment of inhalations will
be found a most important addition to the treatment.
So far as local treatment proves beneficial, we can get
speedier and better results from remedies used in this
way than in any other.

Experience has conclusively proven that the local
affection is dependent upon an impairment of the
blood, and a correct treatment must be directed to this.
Still the local manifestation in the throat, not only
causes much suffering, but if allowed to progress, reacts
and increases the general disease. It may also prove
fatal by destroying the vitality of the structures affected.
or by extending to the larynx.

This local affection is not to be arrested by the use of
escharotics, as was first attempted, and many physicians
now only use a solution of Chlorate of Potash as a gargle, or something equally simple and mild.

An inhalation of a saturated solution of Chlorate of
Potash (with an atomizer), or a solution of Sesquicarbo-
nate of Ammonia or of Sulphite of Soda, will not only
give temporary ease, but materially assists in the cure.
The employment of a decoction of tansy, or of German
Chamomile, Baptisia Tinctoria, or of simple vinegar and
water, vaporized in the common manner, prove useful.
Permanganate of Potash in the proportion of grs. ij to
grs. xv, to the ounce of water, is a good application.
Formula 22 is a very good one, as is No. 19. The importance of these means in cases of children too young
to gargle, will readily be seen.

If the disease extends to the larynx, these means are
of still greater importance. Many physicians now consider the case nearly if not quite hopeless, but I am
satisfied that if inhalations are properly employed,
quite a large number of these cases may be saved.
Even the employment of the vapor of vinegar, or of
equal parts of vinegar and water, is attended with
marked benèfit. The first case of diphtheritic laryngitis in my practice, occurred in the winter of 1860.
The patient, a young lady aged 19, was attacked
with diphtheria on Friday evening, and by midnight
on Saturday, it seemed almost impossible for her to live,
respiration was so difficult. The common means employed had been continuously used, but so far without
benefit. Inhalations of the vapor of mild cider vinegar were commenced and cautiously employed, and by
Monday evening she was out of danger.

In a second case, Mrs. M——, aged 36 years, the
diphtheritic laryngitis commenced on the fourth day of
the disease. Its progress was very rapid, and the treatment pursued seemed to give but little relief. Becoming much worse at night, I was sent for, but being away

from home, the nurse suggested the use of the vapor of
a decoction of tansy. When I reached the house, she
had been using it about an hour, with some relief: it
was continued, and with the internal use of nauseants,
she was breathing freely in twenty-four hours.

These were marked cases, and the action of inhalations were so decidedly beneficial that I continued their
use in every case. The infusions first named under the
head of relaxants, or under the head of tonics, acidulated with vinegar, will be found most generally available. But I have no doubt that the use of lime water
as recommended for pseudo-membranous croup will be
found equally useful in this disease.

SCARLATINA.

In scarlatina and anginosa maligna, the disease of the
throat sometimes assumes great importance, and demands much care in its treatment. If the patient is of
that age that gargles or other local applications can be
thoroughly employed, we may get along well enough,
but in young children we will find an addition to our
therapeutic resources necessary. This addition I think
will be found in the judicious use of inhalations.

Those that have given the best results in my practice, are an inhalation of vinegar and water, of an infusion of Baptisia or Chamomile, or a solution of Chlorate of Potash, or Hydrochlorate of Ammonia. Not
only does the employment of inhalations relieve the disease of the throat, but it quiets that restlessness and irritability of the little sufferer, that proves so exhaustive,
and renders it so much more difficult to manage.

MEASLES.

The irritation of the respiratory mucous membrane, causing the harrassing cough, is the most troublesome part of this disease. It is not only troublesome, but many times is not amenable to the usual cough remedies, and will continue until some serious lesion of the respiratory apparatus ensues. Leaving out the Drosera, which I regard as almost a specific in many of these cases, I prefer to treat this bronchial irritation with inhalations.

Formulas 9 and 12 are very good in these cases. Or we may use either of those from 1 to 6, or the sedative formula from the Consumption Hospital. Not unfrequently an infusion of chamomile, or vinegar and water, if used during the eruption, will be all that is required.

If an irritation of the bronchii and harrassing cough continues afterward, Formulas No. 10 and 11 will be found useful. Or if the secretion is tenacious, and expectoration difficult, they may be alternated with lime water. An infusion of the common red clover taken internally, and used as an inhalation, sometimes gives speedy relief.

PERTUSSIS.

Hooping cough is now treated by most physicians by empirical remedies, or as some would say by *specifics*. The employment of Belladonna, Nitric Acid, Drosera, or Trifolium-in-Fœno, rarely fails of giving the necessary relief, though we may be unable to tell how they act. The last remedy deserves especial mention, as it is simple, common, and very efficient. Take of common red clover dried, a sufficient quantity, cover it with boiling water, and after it has stood two hours, strain with pressure and sweeten. I usually give it in doses of a teaspoonful to a tablespoonful every hour or two.

Inhalations are palliative only, and their influence brief. Still there are some cases in which this influence is desirable. I would place more dependence in Belladonna than any other agent, used as follows—℞ Tincture of Belladonna, ʒij ; Alum, ʒij ; Rose Water, ℥vj : M. Used in cases where there was relaxation of the mucous membrane, and secretion was abundant. If the mucous membrane is dry, with redness and dryness of the throat and fauces, I would change it thus—℞ Tincture of Belladonna, ʒij ; Nitric Acid, gtt. xx ; Water, ℥ij. One or two drachms of these may be used at a time, and repeated as often as necessary. Formula No. 11 is a good inhalation also.

CROUP.

This means of treatment is employed with decided advantage in croup, in fact, in some cases, I place much reliance upon it. Spasmodic and the milder form of mucous croup is readily treated by the common means, though even here, the vapor of water, or of water and vinegar will be found of assistance.

For ten years past I have never treated a severe case of mucous or membranous croup, without making inhalations of vapor an important means. It allays the irritation and produces relaxation of the intrinsic muscles of the larynx, and thus lessens the difficulty of breathing. And increasing secretion, it promotes expectoration in the mucous variety, and loosens the pseudo membrane in the other. An infusion of Hops, of Chamomile, or of Tansy, acidulated with vinegar, may be employed instead of water.

In pseudo-membranous croup, a late writer recommended the employment of Lime Water as an inhalation, claiming that it may be regarded almost as a specific in this serious disease. I have had only an opportunity of testing it in one case, but this was a very marked one. Finding that the inhalation gave relief, it was continued without internal treatment other than small doses of Veratrum, and the application of cloths wrung out of hot water externally. I have used it in one case of mucous croup with like advantage. I employed Lime Water of full strength, using it every hour at first, and for fifteen minutes at a time, then every two or three hours. This remedy deserves attention, and should be thoroughly tested.

As illustrative of the general impression among physicians that inhalations may be of advantage in these affections, and the imperfect knowledge on this subject, I will make two or three quotations from the recent work of Dr. Prosser James on "Sore Throat." He says: "The contact of watery vapor with inflamed mucous membrane is very soothing. It cuts short the congestive stage of the disease, by a supply of moisture, and removes the dryness, heat and itching. It penetrates through the respiratory tract, and often produces more calm than powerful anodynes. Moreover, the vapor may be easily medicated. The common inhalers are more trouble than they are worth. I usually get the patient to breathe through a large cup-sponge which has been dipped in hot water and rapidly squeezed. A less efficacious method is to lean over a

large basin filled with boiling water. The heat may be
kept up by the aid of a spirit lamp."

Speaking of the benefit to be derived from the continuous application of hot water to the throat in croup,
he says: " Dr. Graves considered he saved several patients by this method, but that it is only applicable to
cases seen at the onset." Nor did he propose to give
up other remedies. " I go a step further, and *apply the
hot water to the inside of the larynx*. Inhalations have
already been spoken of in throat disease. It is obvious
that their application in the usual way would be a matter of considerable difficulty in children. Moreover, in
such a disease as croup, the inspiration is so impeded as
of itself to constitute an objection both to the sponge
and the inhaler. It has appeared to me, however, that
if the little patient could be kept in a warm and moist
atmosphere, great benefit must accrue."

CHRONIC LARYNGITIS

It has been claimed that inhalations are the most important means of treatment in this stubborn affection,
and when used properly will effect a cure if this is possible. With such an opinion, I commenced their use,
but have learned by experience that they are only adjuvants to other treatment.

In chronic laryngitis, one of the most important indications is to keep the organ quiescent, and a failure in
this will entail entire failure, no matter what treatment
may be adopted. To the extent then that we can employ inhalations to relieve irritation, and check cough,
they become useful in this respect. The sedative inha-

lations may be employed for the purpose, but we will find Formula 9 and 10 most generally useful.

When the disease has progressed to change of structure and ulceration, the inhalations of Iodine or Nitrate of Silver, as in Formula 14 and 17, may be employed with advantage. But in a majority of cases, the tonic infusions, with Opium, Morphia, or Stramonium will be most generally useful.

Permanganate of Potash, gr. iij to grs. x ; to water, 3ij, has been employed with benefit , when the secretion was muco-purulent.

We have the same guide to their use here that we have in phthisis. If the irritation and cough are relieved, the patient breathing more freely, the inhalation is of advantage, but if, with the arrest of cough, there is a feeling of oppression, they should be suspended.

Dr. Copeland says: " The inspiration of dry or moist vapor has been recommended in *phthisis laryngea* and in other affections of the respiratory apparatus : but those which have been employed, and often too empirically prescribed, have been either too acrid, stimulating or concentrated : and not being confined in their operation to the larynx, but acting on the respiratory surfaces generally, have proved more injurious than beneficial. The action of these cannot be limited : and hence, only those which I have advised above, and which are balsamic, aromatic, emollient, and narcotic. and cannot injure the lungs, should be employed." I think this will be the experience of every one that has employed inhalations to any extent; at least it has been mine.

M. Trousseau, speaking of the inhalation of pulverized fluids, says : " As to the therapeutical effects of pulverized inhalations, in no class of cases is it more

apparent than in the granular condition of the mucous
membrane of the pharynx and larynx, termed 'dys-
phonia clerincorum,' and so common in preachers,
orators, singers and those who habitually overexert
their vocal powers. About a year ago, a woman affected with small pox came into my ward in the Hotel
Dieu. She was suddenly seized with œdema of the
glottis, of so rapid a character as to immediately endan-
ger life. After requesting M. Robert to be in readiness
to perform tracheotomy, I resolved to try the effects of
inhalation, and caused her to breathe a pulverized solution of tannin. So rapid and complete was the relief,
that by the evening, all danger had vanished, and the
operation was dispensed with. Quite recently, I met
with a case of the same affection, œdema glottidis, in a
phthisical patient, who was pregnant and near her time.
By the use of the pulverized inhaler, I was able to pro-
long this woman's life until after the birth of the child,
and attain a result I should otherwise have despaired of
accomplishing. In syphilitic affections of the larynx,
I have been equally happy, and very willingly bear testimony to the efficacy of the invention."

Dr. Maddock reports two cases of chronic laryngitis
in his work, in which he attributes the cure principally
to inhalations. In one, he employed Belladonna, with
the effect of arresting the irritation and cough, and all
symptoms of local disease in seven weeks. In the second, he used inhalations of Iodine and Conium, with
like success.

ACUTE BRONCHITIS.

It has long been noticed that it was essential to have a moist, warm atmosphere in the treatment of this affection, and quite a number of writers place great stress upon it. It is true, their advice is not followed once in a hundred cases; indeed most physicians think they have done their entire duty to their patients when they order their expectorant mixture.

The best observers go further than this, and recommend the use of demulcent and mild narcotic infusions, vaporized for inhalation. Thus Dr. Copland remarks: " The *inhalation of emollient and medicated vapors* is occasionally of much benefit in the sthenic form of the disease, but chiefly in its first and second stages. The vapor arising from a decoction of Marsh Mallows, or from Linseed tea, or from simple warm water, is the best suited to this state; and should be employed from time to time, the *temperature of the apartment* being duly regulated through the treatment, and constantly preserved from about 66° of Fah., to 75°."

And further on he says: " When the expectoration becomes whitish, opaque and thick, the vapor may be rendered more resolvent, by adding a solution of Camphor in vinegar, and extract of Conium or Hyoscyamus to the hot water, or to the emollient infusions now mentioned; and in the asthenic variety, particularly when the difficulty of expectoration and the fits of dyspnœa are distressing, or when the excretion of morbid matter is impeded or suppressed for want of power the medicated vapors and gases recommended in the chronic state of the disease may be tried."

In the first stage of the disease, inhalations of the vapor of water and other mild agents, give more relief

than any other means I have employed. They are
pleasant to the patient, relieve the dryness and sense of
irritation, check the harrassing cough, and speedily re-
establish secretion. Thus they accomplish that for
which *nauseants* are usually employed, and in one-half
or one fourth of the time, and without the decidedly
unpleasant symptoms that attend the action of that
class of remedies. An inhalation of the vapor of water sometimes answers every purpose. It may be used
as often as necessary, by any of the methods named, or
by the employment of Seigle's apparatus, which I think
decidedly the best.

In some cases, an infusion of Hops, or of Poppy
heads, seems to act better than the water alone. An
infusion of Chamomile flowers, or a small portion of
the tincture added to water will also prove useful. In
some cases I have used Tincture Aconite, f3j to water
f3iv. If the circulation is somewhat feeble, I prefer
Belladonna, or use it in conjunction with aconite in the
proportion of f3ss to f3j, to water f3iv. In some cases
the patient suffers greatly from difficult breathing and
oppression, the result of contraction of the muscular
fibres of the bronchii. Here I use Stramonium in the
proportion of—℞ Tincture of Stramonium, f3ss:
water, f3iv.

If the patient's tongue is red, dry and slick, we will
find it of advantage to acidulate the inhalation either
by the addition of vinegar or Hydrochloric Acid. I
use the last in the proportion of Dilute Hydrochloric
Acid, 3ss to water, or one of the infusions named, f3iv.
But if the tongue is broad, white and pasty, I should
employ Lime Water, of full strength, or diluted with
one part of water. With these inhalation, I employ the
direct sedatives, as—℞ Tincture of Veratrum, f3j ;

Tincture of Aconite, gtt. xx; Water fζiv : M. Give a
teaspoonful every hour until the fever abates, using the
sponge bath and hot foot bath, and afterward every
two or three hours. As a local application to the chest
I prefer the mush poultice.

It is very rarely that other means than this will be
needed, and the treatment is pleasant and the cure
speedy. But if the disease progresses, and the expectoration becomes abundant and muco-purulent, we may
then employ the stimulant inhalations heretofore mentioned,—the Acetous Tinctures of Lobelia and of
Opium, Formula 9 and 12, or of Chlorinated Soda,
Formula 19.

ASTHENIC BRONCHITIS.

In this affection we can obtain more decided results
from remedies employed in inhalation, than in any other
way. Of course we would not withhold Quinine, Iron
and stimulants, as these are important parts of the
treatment. So little are inhalations thought of, I have
not unfrequently been asked, whether they were better
than a tonic and stimulant course of treatment in such
affections. It must be understood, in order to get a
fair idea of their value, that they do not replace other
remedies, but are always to be regarded as additional
means. Or if they replace any agents, it is that indirect class termed expectorants.

In this disease, inhalations of a stimulant character
are most generally employed, though some others will
be found equally useful. Thus I have obtained most
marked benefit from the use of Belladonna and Stramonium, as in the following formula—R Tincture of

Belladonna, f̃ij; Dilute Acetic Acid, f̃ij; water, f̃iv; Mix.—℞ Tincture of Stramonium, f̃ss; Dilute Acetic Acid, f̃ij; water f̃iv : Mix. I order an inhalation of from two drachms to half an ounce, every two hours with an *atomizer*, or vaporized in the common manner. If it produces headache or dizziness, lessen the strength of the solution. An inhalation of the Tincture of Hamamelis, f̃ss to water f̃iv, answers a good purpose.

In some cases we find Formula 9 and 12 very good remedies, the Lobelia and Opium acting as powerful stimulants to the relaxed mucous membrane. The turpentine inhalation, Formula No. 15, answers a very good purpose when secretion is profuse, as will the solution of Permanganate of potash, Formula No. 18, or Chlorinated Soda, No. 19. In other cases the astringents may be employed, as of Tannic Acid in Formula No. 23, or Alum in Formula 24. Or the resins and balsamic agents may be used as named under the head of chronic bronchitis.

PULMONARY APOPLEXY.

We use this name for want of a better, as it is generally understood to express great congestion of the structure of the lungs, with effusion into its parenchyma, into the air-cells and bronchial tubes. In some cases there is hemorrhage, but in more a profuse secretion of a rusty, brownish, or dirty sputa.

These are rare cases, but very difficult to manage. and not unfrequently prove fatal under the ordinary treatment. I adopt the following plan. Give internally,—℞. Tincture of Lobelia, Compound Tincture of Lavender, aa f̃j : Simple Sirup, f̃ij : Mix. Give in doses of one teaspoonful every half hour at first, then every one and two hours. Use an inhalation of Cider Vinegar, at first very slowly, increasing the quantity of the vapor, as the patient can bear it. In using Seigle's apparatus, I order—℞. Tincture of Stramonium, f℥j; Dilute Acetic Acid, f℥vj: Mix. This treatment has proven so satisfactory, that I feel that I can hardly recommend it too strongly. Within the last week I have had such a case, the symptoms being so severe within six hours, that no person who saw him supposed he could live until night. Expectoration was so abundant, that it required every effort to free the tubes from the rusty sanious material. Yet this man was relieved in twelve hours, and was up on the fourth day.

CHRONIC BRONCHITIS.

Inhalations have been used to a greater extent in this disease than in most others, at least we find more written upon it. As the disease is one involving change of structure and function of a part readily reached by medicated vapors, we can easily see why it will be amenable to this plan of treatment. Still, as in preceding cases, we must not neglect the equally if not more important object, of improving the general health by the judicious use of bitter tonics, Iron, Hypophosphites, Cod-liver Oil, etc. All physicians who are successful in the treatment of chronic disease, especially diseases of the respiratory organs, recognize this fact, *that the local affection is the more readily cured, the better the general health is.* To this rule there are no exceptions, hence the importance we attach to a general tonic and restorative treatment.

Dr. Copland seems to have had a very clear conception of what might be accomplished by the employment of inhalations in this affection, and employed them to some advantage; but from the imperfection of the apparatus, they were not used to any considerable extent. It will be instructive to read what this author says, upon the subject of inhalations.

"Notwithstanding the unsuccessful attempts of Beddoes to revive the practice, by employing the elementary and permanently elastic gases, but according to views too exclusively chemical, the practice of inhalation has long been neglected or undeservedly fallen into the hands of empirics. Very recently, however, it has been brought again into notice by M. Gannal, Mr. Murray, and Sir C. Scudamore; and *Chlorine Gas*, and fumes of *Iodine*, and watery vapor holding in solution various *narcotics*, have been recommended to be inhaled. I have tried those substances in a few cases of chronic bronchitis; but in not more than two or three cases of tubercular phthisis. The Chlorine was used in so diluted a state as not to excite irritation or cough. The Sulphuret of Iodine, and the *Liquor Potassii Iodidi concentratus* were also employed; one or two drachms of the latter being added to about a pint of water, at the temperature of 130°, and the fumes inhaled for ten or twelve minutes, twice or thrice daily. The tinctures or extracts of Hyoscyamus and Conium, with Camphor, added to water at about the above temperature, were likewise made trial of; and although the cases have been few in which these substances have been thus used by me, yet sufficient evidence of advantage has been furnished to warrant the recommendation of them in this state of the disease.

"*Inhalations*, also, of the fumes of the *balsams*, of the

ⁿⁿ888888888888888888888888888888888888888888888888888888888888888888888888888888888888888888888888888888

terebinthinates, of the odoriferous *resins,* &c., are evidently, from what I have seen of their effects, of much service in the chronic forms of bronchitis: and I believe that they have fallen into disuse, from having been inhaled as they arise in a column or current from the substances yielding them, and before they have been sufficiently diffused in the air. When thus employed, they not only occasion too great excitement of the bronchial surface, but also intercept an equal portion of respirable air, and thereby interfere with the already sufficiently impeded function of respiration. M. Nysten has shown (*Dict. des. Scien. Med.* t. xvii. p. 143,) that ammoniacal and other stimulating fumes, when inhaled into the lungs in too concentrated a state, produce most acute inflammation of the air-tubes, generally terminating in death; and has referred to a case in which he observed this result from an incautious trial of this practice. I conceive, therefore, that the vapors emitted by the more fluid Balsams, Terebinthinates, the Resins, Camphor, Vinegar, etc., and from Chlorine and the preparations of Iodine, should be more diluted by admixture with the atmosphere, previously to being inhaled, than they usually are. According to this view, I have directed them to be diffused in the air of the patient's apartment, regulating the quantity of the fumes, the continuance of the process, and the frequency of its repetition, by the effects produced on the cough, on the quantity and state of the sputa, and on the respiration. The objects had in view have been gradually to diminish the quantity of the sputum, by changing the action of the vessels secreting it; without exciting cough, or increasing the tightness of the chest, or otherwise disordering respiration. From this it will appear, that the prolonged respiration of air

5

containing a weak dose of medicated fumes or vapors, is to be preferred to a short inhalation of them in their more concentrated states. The want of success which Dr. Hastings and others have experienced, evidently has been partly owing to the mode of administering them, and partly to having prescribed them inappropriately. When the patient complains of acute pain in any part of the chest, as in some of Dr. Hasting's cases, they are as likely to be mischievous as beneficial. Where benefit has been obtained, it will be found that it was when the fumes of the more stimulating of those substances were diffused, in moderate quantity, in the air of the patient's apartments; or when he passed, at several periods daily, some time in a room moderately charged with the vapor or fumes of the substance or substances selected for use."

In my practice I very rarely employ the sedative inhalations first named, unless there has been great irritation; in such cases the Formulas from 1 to 6 may be used, or the sedative Formula of the Consumptive Hospital, No's 26 and 27. Using Seigle's Apparatus, I employ an infusion of Lobelia Herb or Verbascum, or Hops, if there is dryness of the mucous membrane. In some cases we may alternate these with Formula No. 9, or No. 11, either of which will prove useful.

If the secretion is tenacious and raised with difficulty, causing hard and prolonged coughing, I employ an inhalation of Lime Water with Stramonium, Belladonna, or Morphia, as may seem indicated by the particular case. Occasionally we will find the Belladonna and Stramonium to act better, as in the Formulas under the head of Asthenic Bronchitis.

If expectoration is free but not excessive, with a feeling of oppression and debility, as regards the respira-

tory function, we will find the Belladonna, alternated
with the Chlorine, as in the Formulas No. 19 and 29,
very good. Iodine, very dilute, may be inhaled with
Conium or Belladonna. But in some of these cases,
Opium will give the best results, as in Formula No. 9,
or associated with some tonic infusion.

Where secretion is profuse, we may employ the stimulant inhalations, or the astringents. I have obtained
very marked benefit from the use of Iodine, and from
the Permanganate of Potash, and in these cases both
Iodine and Chlorine are highly recommended by those
who have made most use of them, Chlorine Water,
used with Seigle's Apparatus, would likely be the best
form. Of the astringent inhalations, Formulas No. 23
and 24, may be used with advantage, or the Acetous
Tincture of Opium may be added to either. The ter-
ebinthinate inhalations are used with advantage in these
cases, and occasionally we derive benefit from the Gum
Resins, or Balsams.

ASTHMA.

As a palliative, inhalations will be found more generally available than other methods of using medicine,
being prompt in their action, and much pleasanter for
the patient. Even an inhalation of the vapor of water
gives relief in many cases. I have employed the Tincture of Stramonium in the proportion of f℥ss. to water
f℥iv: with very marked advantage. Belladonna may
be used in those cases in which there is relaxation of
the bronchial tubes. I usually order it in the following
proportion—℞. Tincture of Belladonna, f℥j. to f℥ij

Water, f℥iv : Mix.   Use half an ounce for an inhalation.

In some cases I have used an infusion of Lobelia, and in others of Verbascum, or the Tinctures may be added to water and employed in the same way.

The use of inhalations of Chloroform and Æther as palliatives are well known, and practiced to a considerable extent.   They may be associated with the means just named.   The sedative inhalations from Nos. 1 to 6 may also be used occasionally.

In those cases where the tongue is broad, pale, and pasty, I have obtained good results from alkaline inhalations.   The Lime Water, heretofore named, answers an excellent purpose, or we may use the Chlorinated Soda, Formula 19.   Occasionally the Permanganate of Potash, will be found useful, as in Formula 18, this is especially useful when secretion is free.

When the secretion is abundant, and the bronchial tubes relaxed, as in that variety termed *Humoral Asthma*, the remedies recommended for Asthenic Bronchitis will be available.

In describing the treatment of Asthma, Dr. Copland thus writes of inhalations:

" Next and, perhaps, equal to smoking, is the inhalation of simply emollient or of medicated vapors into the lungs.   This method of treatment was recommended by Cælius, Aurelianus, Alberti, Mudge, Beddoes, Thilenius, Zallony, Hufeland, Crichton, Forbes, Gannal, Scudamore, and Murray.   It is chiefly indicated during the paroxysm, or shortly before its accession.   The vapors arising from pouring boiling water upon Camphor, any one of the Narcotic Extracts or Tinctures, or the balsams, are of great advantage when properly managed. Thus the vapor from a pint of boiling water poured upon half an ounce of balsam of tolu; or that from a Solution of Camphor, Balsam of Tolu, and Extract of Lettuce, or of Conium, in Sulphuric Æther; or the fumes proceeding from Camphor, Hyoscyamus, and Aromatic Vinegar, mixed together, and quickened by the addition of some boiling water, may be employed. A Solution of Balsam of Tolu in Sulphuric Æther, the vapor of boiling Tar diffused in the air of the patient's chamber, Chlorine Gas much diluted with common air, and various other medicated vapors may be tried; but these act chiefly by removing the viscid phlegm which collects in the Bronchi, and by exciting the extreme exhaling vessels. I have prescribed the vapor of the *Sulphuret of Iodine* in two cases: in one of Spasmodic Asthma, with no benefit; and in one of Humoral Asthma, with only temporary advantage. Sir C. Scudamore recommends this Formula for the inhalation of Iodine—R. Iodinii gr. viij; Potassii Iodidi gr. v; Alcoholis ʒss; Aquæ Destil. ʒvss: M. Fiat Mistura. To this he adds Tincture of Conium. But his directions as to quantity and mode of inhalation are, notwithstanding several attempts to unravel them, perfectly beyond my powers. I believe, however, that portions only of the above mixture should be employed for each inhalation. But the observing practitioner will generally be able to apportion the quantity, as well as to direct the particular materials, for inhalation, according to the peculiarities of the base; bearing in recollection that the combination of narcotic and anodyne vapors with the volatile fumes and gases will generally be of more service in Asthma than the use of individual substances belonging to one only of these classes of medicines; and that the more irritating sub-

stances of this description, such as Iodine, Chlorine,
and Tar vapor; should be ventured upon only in a very
weak or diluted state."

PNEUMONIA

The opinions of the profession in regard to the treatment of inflammation of the lungs, has undergone a
decided revolution in the last twenty years. The old
fashioned antiphlogistic treatment has been proven
worse than useless, and those *heroic* means, blood-letting,
mercurials, and antimony, with their associates, are
buried so deep that we trust there will never be a resur-
rection for them. In theory we may talk about a
change of type in disease to account for the change of
treatment, and cover the shortcomings of the old practice. But admitting that in olden time diseases were
of a more sthenic character, it was only that much better for the patient, as they were more readily managed
by simple means.

So entirely futile were the old means, that many
physicians, and some of the best writers in England,
almost or quite discard medicine and rely exclusively
on the mush poultice, and diet and rest.

I would not go this far. Thus I find that in the proportion that I control the circulation by the use of the
special sedatives, I control the inflammatory process.
And that this is aided by such means as relieve irritation of the bronchial mucous membrane. Thus I order
Veratrum in the usual doses, with a mush poultice to
the chest. In place of giving the usual expectorant
remedies for the relief of the cough, I think I obtain
better results from the use of inhalations.

At first I employ the vapor of water, or of an infusion of Hops or Chamomile, as recommended in acute bronchitis. When secretion is established, I acidulate these infusions with cider vinegar. These generally are all that is required, but if the case proves stubborn, and the secretion profuse, or the stage of grey hepatization is reached, I prefer the solution of Permanganate of Potash, or of Chlorinated Soda. It is hardly worth while to name other remedies, as what was stated in the consideration of acute bronchitis is equally applicable here.

PHTHISIS PULMONALIS.

Consumption is now regarded by the better informed physician as a disease of nutrition, and medication is directed to this end rather than to the lungs. Occuring in persons of feeble vitality, either congenital or acquired, we observe a still further depression previous to, or with the development of the symptoms of phthisis. Two causes may be said to exist at the same time, an imperfect elaboration of the blood, and defective secretion. With imperfect or feeble vitality, the blood is always of lower organization than in persons in robust health, and there is always more or less imperfectly formed material that must be removed by way of the excretory organs. We will therefore find them doing a larger amount of labor than in others. In addition, the tissues do not possess so high a grade of vitality and give way much easier. If, therefore, there is such further impairment of vitality in these cases as to lower the organization of the blood, we shall have the material of tubercle (imperfectly formed albumen), in excess. If the excretory organs are active, this material is removed by way of the skin, kidneys and bowels, and no deposit occurs. But should they fail, the material circulates in the blood as a foreign body, and will be thrown out as tubercle, whenever an irritation with determination of blood is set up.

Not only so, but it is clearly proven that the broken down elements of the tissues is a source of tuberculous material, in persons of feeble vitality. If this be the case, and it is conclusive to my mind, we cannot but see the imperative necessity of keeping the secretions free for its removal.

The indications of treatment are therefore three in number.

1. To place the stomach and digestive canal in good condition, give the patient an appetite and power of digestion, and such restoratives as serve to make a better quality of blood.

2. To restore and keep active, secretion from the skin, kidneys and bowels, and thus remove material that cannot be used for nutrition of texture, but will be deposited as tubercle if not removed in this way.

3. To check irritation of the lungs, controling the cough and determination of blood, and thereby checking the deposit therein until the material may be removed in other ways.

It seems to me that these indications are very clearly deducible from the pathology of the disease, and experience has proven to me, that the practice based upon them is quite successful. Indeed, if they are fully appreciated, it is hardly worth while to speak of special medicines, as the proper ones will immediately suggest themselves. We have, unfortunately, no specifics for consumption, and as far as treatment proves useful, it must be based upon the principles above laid down.

One great stumbling block in the way of correct treatment, is the fear of increasing the debility by the use of means to establish secretion, and place the stomach and bowels in good condition. It is true, that all debilitating treatment should be avoided, but with the proper remedies to fulfill these indications no debility will result. Thus I do not hesitate to give a thorough emetic, where the stomach is in an atonic condition, and there are morbid accumulations and increased secretion of mucus, and repeat it if necessary. Or, to give a mild cathartic of Podophyllin and Leptandrin, with Hydrastin and Quinine, to get normal activity of the upper intestine, and promote secretion. Stimulate the skin to action by baths, alkaline, tonic, stimulant, astringent, oleaginous, or vapor, as the case may need.

In some cases I have trusted to a great extent to Iron, frequently using the muriated tincture, one part to three of Glycerine, in doses of a teaspoonful four times a day. The Collinsonia Canadensis, is an excellent remedy, fulfilling the double indication of a tonic, and a remedy to control irritation of the lungs and cough. The Hypophosphites will be found useful where there is much excitation of the nervous system, and where there is a languid circulation of blood. The compound syrup is a good preparation, or we may use the Hypophosphites of Lime and Soda in alternation. In place of these, the Tincture of Phosphorus, $\mathfrak{z}$j, to Simple Syrup, $\mathfrak{z}$iv, may be given in teaspoonful doses every four hours. Cod-liver oil proves useful when it is readily taken and digested, but if it nauseates and impairs the appetite, it should be omitted. Small quantities of Bourbon Whisky are occasionally used with advantage; but should

7

never be employed as a substitute for food, or to such
an extent as to impair the appetite.

In the fulfillment of the third indication we have the
use of inhalations in the treatment of this disease.
And here I wish to be distinctly understood as not
claiming that they will supplant the use of the ordinary
means in all cases, but only that they may be used in
addition to them, and in some cases entirely replace
them.

The agents employed in inhalation will depend to a
considerable extent upon the amount and kind of secretion from the bronchial tubes, and the ease and difficulty with which it is raised. Again, we have to take
into consideration the state of the mucous membrane,
as regards tonicity and circulation, and lastly the general health of the patient.

It may be laid down as a general rule, to which there
are few if any exceptions, that an inhalation which
relieves the cough, and gives ease and freedom in respiration, cannot but prove beneficial. And, on the contrary, any remedy that unduly excites, causes determination of blood, or gives rise to a feeling of fulness and
oppression in the chest, will always prove detrimental.
Bearing this in mind we will rarely cause harm by the
use of these means, as I have not unfrequently wit-
nessed in the treatment of those who make inhalation
a specialty.

If there is dryness of the air passages, I order an inhalation of the vapor of water, slightly acidulated or
rendered alkaline, according to the condition of the
tongue. If the tongue is red, and inclined to dryness,
sometimes slick, I always acidulate the inhalation,
either with vinegar, or some other of the vegetable
acids. If the tongue is pale, broad, and covered with a

pasty coat, I always render them alkaline by the addition of soda or of lime. I attach great importance to these suggestions, as I believe them founded upon an unchanging therapeutic law, which is equally applicable to the administration of remedies by mouth.

An infusion of Hops, of Chamomile, of Marsh Mallow, or even of some of the simple bitter tonics will be found useful here. If there is great irritation and harrassing cough, and these means do not seem to relieve it, the sedative or narcotic inhalations may be alternated with them, or added in proper proportion to them.

Of the sedative inhalations, I prefer Formulas No. 3, 5, 6 and 27, as they allay irritation and cough, without producing oppression. These inhalations will be applicable to those cases in which secretion is moderate, and expectoration not difficult.

If the sputa is tenacious, and raised with difficulty, we obtain the best results from an infusion of Lobelia alone, or with Opium as in Formula No. 9. If there is difficult breathing from contraction of the bronchii, the Stramonium as in Formula No. 11, will prove better. In place of these we may not unfrequently obtain better results from Lime Water, with the addition of a solution of Acetate of Morphine.

If the secretion is abundant, we employ stimulants and astringents. It is in these cases that we obtain benefit from the use of Iodine and Chlorine. We may employ Iodine as in Formulas 33 and 14, with the apparatus represented in Figures 1 and 2. Or with the atomizer as in Formula 13.

In using Iodine, we commence with small quantities, and I prefer it with the vapor of water, as in using it with the atomizer. Occasionally we prescribe Bella-

donna or Stramonium with it, or Acetate of Morphia, or
other of the sedatives.

The treatment of pulmonary consumption by Iodine
is very frequent in Belgium, and has been especially
recommended by M. Chartroule. Under his directions
twenty-eight patients in the hospital were treated by
the inhalation of the vapor of pure Iodine, and of this
number only eleven could be said to derive no benefit
from the treatment. In these unsuccessful cases, the
pulmonary lesions were not modified, but still the
symptoms were not aggravated in any case. In oppo-
sition to the statement that Iodine vapor produced hemoptysis, it was found that pulmonary hemorrhage
ceased more rapidly under this kind of treatment than
under other plans which are more generally employed.
Seventeen patients derived positive benefit from the
Iodine treatment, and this improvement was observed
not only in relation to the general symptoms, but also
to the pulmonary lesion itself, as was proved by per-
cussion and auscultation. Out of the seventeen patients, four might be considered as actually cured. One
of the cases of cure is the following:—A youth, six-
teen years of age, entered the hospital in such an alarm-
ing condition that at first the physicians hesitated to
submit him to the Iodine inhalations. He was in a
state of great emaciation, and his skin was almost constantly covered with profuse perspiration; he had
diarrhœa, which had lasted for two months, and he had
repeatedly suffered from hemoptysis. There were very
extensive indurations in the lungs, and at the apex of
the right lung there was a cavity of some size, as was
shown by very obvious gargouillement. The expectoration was also characteristic. After resting a few
days, this young man was subjected to the Iodine inha-

lations, and all the symptoms which had appeared so serious were soon modified in a most remarkable manner. The general symptoms disappeared first, and the body recovered its plumpness with great rapidity. The perspiration, diarrhœa, fever, cough, and expectoration were soon relieved or removed, and six weeks after admission into the hospital the patient went out quite well.

I prefer Chlorine in the form of Chlorine Water, used with the atomizer. That is best, that is prepared by passing the Chlorine disengaged by the action of Hydrochloric Acid and Binoxide of Manganese, into water to saturation. The Middlesex Formula gives a very good solution of Chlorine, and as it is easily prepared, it will be most generally used—℞. Potasssæ Chloritis, ʒij ; Acidi Hydrochlorici, Aquæ Distillatæ aa. fʒij : Misce. This should be kept in a stoppered bottle in a dark place. In using it from, fʒss to fʒj may be added to fʒj, of water. In some cases it will have to be used much weaker than this.

Turpentine may be employed with advantage in some cases, either alone, or in connection with the sedatives. Tar water has likewise had considerable reputation, and a quite celebrated physician was accustomed to use rum that had been boiled with tar.

In some cases of phthisis, coal oil has been employed with reported advantage as an inhalation. It seems better adapted to those cases in which there is abundant expectoration, with great relaxation of the bronchial mucous membrane. Add to four ounces of Coal Oil, ten grains of Sulphate of Morphia, and use from two to four drachms as an inhalation.

In place of these we may use the astringent inhalations, but as a general rule they will be found to cause

oppression and difficult breathing, and will have to be
abandoned.

In concluding this subject, I will make an extract
from the work of Dr. Maddock, not that it throws any
more light upon the subject, but as showing that authority is not wanting for many of the statements I have
here made:

"Dr. Bodtcher, of Copenhagen, has published some
interesting observations on the efficacy of camphorated
vapors in complaints of the air passages; Raspail has
also strongly recommended the use of Camphor in nervous and spasmodic affections of the air passages, the
patient taking it powdered, as snuff, or respiring its
vapor—small pieces of Camphor being enclosed in a
straw, or quill, the ends closed with cotton-wool, the
tube then placed in the mouth, and the breath drawn
through it. Dr. Harwood speaks highly of the inhalations of Ammonia and Camphor, employed with a temperature of about 100°, which are frequently very useful in relieving distressing symptoms, and in promoting
the cure of some affections of the fauces, the larynx.
and trachea, among which the most common are
hoarseness and loss of voice; their benefit arising either
from acting directly on the part affected, or from com-
municating their influence along a limited extent of
certain nerves of the throat, by a sympathetic action.
On other occasions, by suitably diminishing the stimu-
lus of these inhalations their influence may be safely
extended to more remote parts of the pulmonary nervous system ; and hence in some chronic complaints of
the chest, in elderly persons, much benefit has attended
the addition of a little Ammonia to Ammoniacum, or
other expectorants, with a view to arouse and augment
nervous power in the lungs, in consequence of its

being so far diminished, as to render the removal of phlegm from the air passages very difficult.

"Sir A. Crichton, in an 'Account of some Experiments made with the Vapor of Boiling Tar, in the Cure of Pulmonary Consumption,' after detailing various cases which had come under his treatment, makes the following remarks: 'It must be evident that the Tar fumigation, though most completely successful in some of them, did not produce the same good effects in all; but, on the other hand, the very great relief which every patient experienced at first from it, particularly in the diminution of cough, expectoration and fever, is a fact which ought to encourage us to multiply the trials of this remedy as far as possible. * * * The Tar vapor seems to have healed the ulcers, and removed the inflammation of the tubercles, in the greater number of cases, but I do not believe it produces the absorption of the tubercles themselves. * * * At that period when the cough, expectoration, and hectic fever are greatly subdued by the influence of the Tar fumigation, it seems to me often injudicious to continue it longer, or at least, in so strong a degree as before. Notwithstanding the great power of this means of cure, I never employed it quite alone, but at the same time prescribed internal remedies, such as the nature and urgency of the symptoms seemed to require; but these have been the same as every practical physician has recommended in similar cases.'

"Since the introduction of Tar, by Sir A. Crichton, in the year 1817, numerous trials of it have been made by Lazareto, Hufeland, Dr. Morton, of Philadelphia, Dr. Neumann of Berlin, and many others, all of whom are sceptical of its value as a *curative* agent in pulmonic diseases, but believe it capable of diminishing the night sweats, the expectoration, and hectic fever.

" The mode of using the Tar preparation consists in boiling some common Tar, and adding to each pound from one to two ounces of the Carbonate of Potash, to destroy the Empyreumatic Acid; a small quantity of this is put over a spirit lamp, and by thus disengaging the volatile part of the Tar, which consists of an invisible vapor, the air of the apartment soon becomes impregnated.

" Inhalations of creosote, a preparation of Tar, were extensively tried by Dr. Elliotson a few years ago, but were speedily discontinued as being of little or no value.

" We have employed Tar, Creosote, and that species of Tar termed naphtha, or pyro-acetic spirit, in a great number of cases of pulmonary consumption with the utmost caution and perseverance, and in exact accordance with the directions enforced by their respective advocates as to quantity and quality, but without deriving the beneficial effects which have been attributed to them, and we believe their efficacy is of very limited applicability, and refers only to some few phenomena and effects of the disease.

" Dr. Bennet, in his work entitled Theatrum Tabidorum (chap. De Halituum et Suffituum), records several cases of pulmonary disease successfully treated by inhalations of various gases and watery vapors, with accounts of the fumigating apparatus, and recipes for the remedies.

" Dr. Cottereau, of Paris, has communicated, in the Journal Hebdomadaire, and the Arch. Gen. de Medecine, for 1830, several highly important cases of pulmonary consumption, in which perfect recovery ensued from the use of chlorine inhalations. Mr. John Murray, in a most interesting and able work on pulmonary consumption, has also narrated numerous cases of pulmonic disease which had been cured by the same remedy. Dr. Elliotson, in the Lancet, No. 402, observes in his admirable lectures, that he has seen many cases of tuberculous consumption and diseases of the air passages in which the more distressing symptoms were quickly relieved by the inhalation of chlorine; but hesitates to give a decided opinion of its curative effects until he has made further trials. Dr. Elliotson at the same time remarks, that the medical profession have been much to blame for neglecting the inhalation of various substances, and allowing their patients to die under the old 'jog-trot' system, well established as unsuccessful; and that the duty they owe to themselves and their patients demands that they should not persist in affording alleviations only, when there was the slightest possibility of accomplishing more good than before by any new means. Dr. Elliotson adds, that ' it shows a very narrow mind to set one's face against attempts at improvement; and I, therefore, give credit to all my medical brethren who suggest anything new, and still more to those who make exertion to carry such things into effect.' "

V.

ON THE USE OF ATOMIZED FLUIDS IN OTHER AFFECTIONS.

Though not entirely pertinent to the subject of the work, I cannot let the opportunity pass of saying something with regard to the use of atomized fluids in other diseases, than those of the respiratory organs. What I have to say here, is the result of my own experience, and will have to be corrected and enlarged by the practice of others. I am satisfied, however, that the facts will stand the proof of experiment, and prove a valuable addition to our therapeutic resources.

Whilst we employ the same remedies that we would use as local applications in the same diseases, we obtain a far greater influence by their minute subdivision, and by their equal application to the diseased surface. In many cases, a local application, however carefully applied, only reaches a part of the affected surface, and is very speedily removed by the natural secretion of the part. With the atomized fluid, it penetrates every part, and is continually renewed as long as the operator desires, whether it be five minutes or one hour.

THE EYE.

The applicability of this means of medication in diseases of the eyes will be readily seen. The warmth and moisture relieves for the time that irritability that so frequently prevents the proper application of eye-washes. The fluid reaches every part, is equally distributed, and its application continued as long as desired. If these are facts, and any person may readily test them for himself, it will be seen that this method is a distinct step in advance of the common practice.

If they were no better in the case of the adult, this could not be claimed in treating diseases of the eyes in children. Here we find it almost impossible to use even the most simple collyria with good results, and even when it is applied, the force used, and the resistance of the child, does more harm than the local application does good. But with Seigle's atomizer we have no difficulty whatever, as the soothing influence of the vapor more than counterbalances the effects of the medicine.

ACUTE CONJUNCTIVITIS.—This is the case in which atomized fluids will be found most strikingly beneficial, as the influence of the vapor is emollient and soothing, and relieves of itself, to some extent, the irritation upon which the determination of blood depends. We may employ the vapor of water alone, or we may use any mucilaginous infusion. I generally order—℞ Tincture of Belladonna, f℥j; water, f℥iv: or—℞ Tincture of Belladonna, f℥j; Tincture of Gelseminum, f℥ij; water, ℥iv: or—℞ Aromatic Sulphuric Acid, gtt. xv; Morphia Sul. gr. ij; Aqua, f℥ij. In place of the Tincture of Belladonna, Atropia might be used, but it would have to be very largely diluted.

PURULENT CONJUNCTIVITIS.—In purulent conjunctivitis these same remedies may be employed, the combination of Aromatic Sulphuric Acid and Morphia being especially useful. In other cases we may use an infusion of Hydrastis or Baptisia, or of Chamomile, alternated with the common collyria. A weak solution of Carbonate of Ammonia, or Chlorate of Potash, say from one to five grains to the ounce of water, will prove useful. Or if the discharge is very free, we may use a solution of Permanganate of Potash, one to three grains to the ounce of water. The Acetous Tincture of Opium is sometimes a good remedy in these cases, in the proportion of from f$\mathmedical{3}$ij, to f$\mathmedical{3}$ss, to $\mathmedical{3}$iv of water.

CHRONIC CONJUNCTIVITIS.—In this disease, we find that the ordinary mild collyria may be used with considerable advantage. Thus I have obtained very good results from the use of Belladonna, and the acid Collyrium heretofore named. The employment of the tonic infusions will also prove beneficial. Either of the tonic or astringent formula, heretofore given, may be employed, as in Formulas No. 22, 23, and 24, lessening the Aromatic Sulphuric Acid in the first two to f$\mathmedical{3}$ss.

In granular conjunctivitis, these means are not so available, though here they may be employed as adjuvants to the treatment, as the milder inhalations are generally used.

RHEUMATIC OPHTHALMIA.—In this case Collyria have been generally regarded as useless. But the atomized fluid relieves irritation and soothes the inflamed organ. I generally order—℞ Tincture of Aconite gtt. xx; Tincture of Belladonna, f$\mathmedical{3}$j; water, f$\mathmedical{3}$iv: Mix. The same means might be used in all internal inflammations, or even in iritis, though in the last, the use of a collyrium of Atropia will be found much better.

In other affections of the eyes where Collyria is used, the employment of atomized fluids may be substituted, special means need not be named in these cases, as they do not differ from those commonly employed.

THE EAR.

Of course the employment of the atomizer in diseases of the ear is but limited, being confined entirely to affections of the external auditory meatus. But even this limited action is sometimes available, as no diseases are more painful than those affecting this part.

Acute inflammation of the ear may be more speedily relieved by the use of narcotics in this way, than by any other means. Tinctures of Aconite, Stramonium, Belladonna, or Opium, added to water may be thus employed. I have used ℞.—Tincture Aconite, (root,) f℥ij; Tinct. Stramonium, f℥ss; Water, f℥iijss. Mix.

In *otorrhœa*, we employ similar means to those we use as injections. The Tincture of Muriate of Iron, or solution of Perchloride of iron, f℥ss to Water f℥iv; sometimes answers well. At other times I have used a solution of Permanganate of Potash, grs. ij to grs. v.; to Water f℥j. The astringent and tonic Formula may be employed in the same proportions they are used in diseases of the chest.

OTHER LOCAL APPLICATIONS.

Though it might be employed in any local affection, we will find that its use will be principally confined to the treatment of ulcers. For this purpose it sometimes

proves very valuable, though at others no better, if so
good, as other applications. If used, the usual remedies
will be employed.

It has been used to a limited extent in neuralgic affections, but when any change from the common mode
was necessary, we would find the best means in the use
of the hypodermic syringe.

Some experiments have been made on *Diseases of the
Uterus*, but nothing definite can be said about its application here. Whilst it would seem to be a desirable
method of applying remedies, the disagreebleness of its
employment, and limited action, will prevent its being
used to any considerable extent. In painful affections
of the os uteri, as in ulceration and cancer, and in dys-
menorrhœa, it will doubtless be used to some extent.
The fact that a cylindrical speculum has to be used, and
that the fluid is therefore confined to a very small part,
is an insuperable objection to its use in other cases.

VI.

ON THE EMPLOYMENT OF ATOMIZED FLUIDS AS DISINFECTANTS, AND TO DESTROY ANIMAL AND VEGETABLE MIASMS.

An efficient means of disinfecting sick rooms, hospitals, prisons, ships, etc, has long been desired. We have disinfecting fluids in abundance, but they have never as yet been so applied as to be of much value.

I have determined by experiment that we may remove offensive odors, and destroy the germs of animal miasm, very speedily and with great certainty with the apparatus heretofore described. Thus taking the coal oil atomizer, Fig. 6, I have purified the air of a large and very foul dissecting room in an hour, and that with the use of but one ounce of the Solution of Chlorinated Soda. I have attained the same results by the employment of a solution of Sulphate of Iron, and the Sulphite of Soda. It is done so thoroughly that the room would not be offensive to. any person. And as evidences that the odor was not simply covered up, we find that it is reproduced very slowly and only by renewed processes of decomposition.

With this apparatus, the air of a sick chamber can be thoroughly freed of all bad odors, and what is more to the purpose, the animal miasm upon which the spread of disease so frequently depends, is entirely destroyed. This is of very great importance, as the sick in some affections are continually poisoned by the decomposing emanations from their own bodies. Not only is it of advantage to the sick, but we thus prevent the spread of contagious diseases.

In seasons of epidemic disease, it will prove especially beneficial. The evidence has been growing stronger and stronger, until but few will deny that the spread of contagion depends to a very considerable extent upon an atmosphere already impure from animal and vegetable decomposition. It matters not whether it is typhoid fever, or cholera, or small-pox, measles, or scarlatina, or dysentery, the same rule holds good. During such epidemics, therefore, especial attention should be paid to the disinfecting of all places where decomposition is going on.

When the popular mind becomes alarmed by the approach of an epidemic, as of cholera, there is a spasmodic effort toward cleanliness, which is rather injurious than otherwise, in this—that a much larger and fresh surface is exposed to the action of the air; in fact the dirt is stirred up in place of being removed, and that at the time when it should remain at rest. I do not desire to advocate dirt, but I object decidedly to placing it in the most active condition possible during an epidemic. Do the cleaning before the epidemic comes, but when it is upon us, destroy the products of decomposition and keep it in as quiescent a condition as possible.

The apparatus represented in Fig. 6 will be sufficient

to thoroughly free a dwelling house from all unpleasant odors and animal miasms of which they are the sign. I prefer the use of the solution of Chlorinated Soda, but Chloride of Lime, or Sulphate of Iron may be employed with equal advantage. For more extended use in epidemics, I employ a much larger apparatus, using flexible hose, which can be carried wherever necessary.

It is very certain that the process of absorption is very rapid from the lungs, especially when the agent is in a gaseous form. Air becomes charged with moisture, and, in case of medicated fluids, with the medicine to a certain degree. May we not, therefore, suppose that in some cases agents might be introduced into the circulation with advantage? I have experimented in this line to some extent, but hardly like to express an opinion as yet.

Should this be the fact, it is barely possible that we may thus find an antidote to the blood poisons, as in cholera, typhoid and typhus fevers, the eruptive fevers, etc. I think the subject is worthy of investigation and would be glad to learn of any investigations in this direction.

APPENDIX

ON

Diseases of the Nose and Throat,

BY

WM. BYRD SCUDDER, M. D.

Professor of Ophthalmology and Otology, and Diseases of the Nose and
Throat, in the Eclectic Medical Institute, Cincinnati.

ANATOMY.

In prefacing disease and its treatment by a few remarks on the anatomy of the parts, the reason is offered that the reader is liable to overlook the anatomy, perhaps thinking it too dry, or perhaps, being busy, noting only the disease, the treatment alone being of interest.

But I beg those who may be interested in this short, and, I hope, practical paper on nose and throat diseases, to glance carefully over the anatomy and physiology of the parts, for although this same point is emphasized in regard to other diseases, still more is it necessary here, for it is difficult to treat and impossible to operate without a general idea of the size, structure and position of the various parts in the nasal fossæ. The nose is the special organ of the sense of smell, and while many authors would class this sense as the least important of the five special senses, yet it has functions that the others have not. It aids us in discriminating between good and poor food, and it protects the lungs and many times the life by detecting foul gases and irritating vapors.

This organ of the sense of smell is divided into the *external*, the *nose*, and the *internal*, the *nasal fossæ*.

Of the external, the *nose*, I shall say little, as it seldom is diseased, except from acne, inflammatory conditions of the sebaceous glands, cancerous degeneration and blows and injuries of different kinds.

There are certain physicians paying special attention
to cosmetic work, who claim to make good looking noses
out of poor ones, but I doubt the ability of any one to
change the characteristic low bridge of the Ethiopian to
the high one of the Israelite, or *vice versa* the high to the
low one. However, there comes a time in the lives of
certain people, when by a blow, the bridge of the nose
and its support underneath are crushed and knocked flat,
and in these cases by forcing the nose outward by means
of a strong probe from the inside, the patient may end
with a better shaped nose than he had before the injury.

Leaving now the nose, we next notice the *nasal fossæ*,
the seat of the sense of smell and the beginning of the
respiratory tract, strange, tortuous and highly vascular.

The *nasal fossæ* are two irregular cavities, with their
direction from before backward, beginning at the anterior
nares or nostrils and ending in the pharynx by means of
the posterior nares.

The posterior nares are two oval openings each about
the size of a hickory nut; that is, one inch high by one-
half inch wide.

The *mucous membrane* lining the nasal fossæ was first
described by Schneider, an anatomist, who showed that
the secretion noticed in the nose came from the nasal
mucous membrane, and not from the brain as was supposed previous to that time.

It is therefore called the *Schneiderian* membrane ; it is
also called the *pituitary* membrane.

It is most important to note that the nasal mucous
membrane extends in many directions; it is continuous
through the nasal duct with the conjunctiva of the
eyes, through the various openings into the sinuses or
pneumatic chambers of the skull, such as the frontal,
sphenoidal, ethmoidal and the sinuses of the superior

maxillary bone called the antra of Highmore: It is also
continuous through the eustachian tubes to the middle
ear and the mastoid, as is only too frequently apparent
by the large amount of ear disease due to this particular
continuity of structure.

The nasal fossæ are separated by the *septum*, a thin
partition which forms the inner boundary of each. The
septum is formed in front by a large triangular cartilage,
above by the perpendicular process of the ethmoid and
posteriorly by the vomer.

The *outer wall* of each fossæ, slanting downward and
outward, is formed by the vertical plate of the palate bone
and the nasal surfaces of the superior maxillary.

The *floor* of the nasal fossæ is formed by the palatal
processes of the same bones, viz: the palate and the
superior maxillary.

The strange feature of each fossæ is the projection of
three horizontal plates of bone that spring from the outer
wall, delicate and shell-like in form, called the turbinated
bones.

The superior turbinate, the smallest of the three, hangs
downward from the roof; the middle and inferior turbinates have their attachment to the outer wall, and curl
inwards and downwards toward the septum.

The projecting bones with their coverings form spaces
between themselves called meatuses; thus the space between the superior and middle turbinated bone is called
the *superior meatus*, the space between the middle and
the inferior turbinated bone is called the *middle meatus*,
and between the inferior turbinate and the floor of the
nose is the *inferior meatus*.

Connected with the nasal fossæ are three sets of air
cavities or sinuses.   The frontal sinus with its canal,
the *infundibulum*, opening into the middle meatus, the

sphenoidal sinuses opening into the superior meatus, and
the antra of Highmore, or maxillary sinuses, situated
in the body of the superior maxillary bone and opening
into the middle meatus.

Into the inferior meatus opens the nasal duct with its
imperfect valve formed by the mucous membrane.

STRUCTURE OF THE TURBINATES.

In noticing the nasal fossæ of a skull from which all
the soft tissues have been removed, the nasal space seems
to be ample and the idea of occlusion of these passages,
called *stenosis*, would almost seem impossible, but the
mucous membrane in this vicinity is thick and highly
vascular, and one, who has not specially looked up this
point, can have no idea of the *thickness* and toughness
the membrane often assumes.

The turbinated bones themselves are very thin and
delicate, and form a field easy of access to ulcerative
processes, to syphilitic invasion and to necrosis.

The first coat is the periostium, a trifle thicker, but
similar to that of the other bones of the body.

The next covering, and that which is *most practical* to
know, is the vascular layer, or, as it has been termed,
the *corpora cavernosa* of the nose.

The corpora-cavernosa, especially of the two lower
turbinates, and, to a certain extent of the septum, consists
of *a dense network or plexus of arteries and veins*, under
direct vaso-motor influence of the *sympathetic*, supplied
by the nasal branches of the spheno-palatine ganglion.

It resembles more than any other tissue in the body
the corpora cavernosa of the penis, and under any form
of irritation, as hot dry air, irritating gases or dust, and
at times even in sexual or menstrual excitement, there is

the *vascular engorgement* which causes the great swelling of the turbinates, thereby producing the *stenosis* which is such a prevalent trouble in nasal disorders.

The next covering is the mucous membrane, very similar to mucous membrane elsewhere, but a little thicker, and covered by an epithelial coat, the cells varying according to location.

VASCULAR SUPPLY.

The roof of the nose is supplied by the anterior and posterior ethmoidal arteries, branches of the ophthalmic artery; these vessels also supply the frontal sinus and the ethmoidal cells.

The chief and most important point in the vascular distribution of the nasal fossæ is the dense net-work of vessels forming the sub-mucous layer or the corpus cavernosa, *especially of the two lower turbinates and the septum*, fed by the naso-palatine artery under direct vaso-motor influence of the nasal branch of the spheno-palatine ganglion.

NERVOUS SUPPLY.

The olfactory nerve is distributed to the roof of the nose, all of the superior turbinated and to the upper third, both of the septum and the middle turbinated bone.

Part of the septum and the middle and inferior turbinated bones are supplied by the nasal branch of the spheno-palatine ganglion; the vidian nerve supplies the upper and back part of the septum, and the superior turbinated bone, while the anterior part of the inferior turbinated bone and the floor of the nose are supplied by the nasal branch of the fifth pair of nerves.

7

The nasal mucous membrane throughout is very sensitive and will not admit of the same astringent treatment as mucous membrane elsewhere.

PHYSIOLOGY.

Speaking practically and also from a physiological standpoint, the nose has two distinct functions, viz: *olfaction* and *respiration*, or an olfactory and respiratory region. The olfactory region is essentially the upper part or the roof of the nose; or, in other words, the seat of the distribution of the olfactory nerve. In ordinary respiration, the air contained in this region is more or less residual air, and it is principally in order to acquire air containing odoriferous particles that we notice a person sniff the air; or, in other words, make a forced quick inspiration to throw the new air into this region, when trying to scent some particular odor.

As regards the sense of smell, there is no doubt but that odorous bodies cast off small particles, which, floating in the air, adhere to the mucus overlying the olfactory nerve, and by some means convey to the brain the sensation of smell.

There are numerous theories regarding the sense of smell, a chemical theory, one of pigmentation, another of irritation, still another of molecular vibration, but these will be of no particular interest here.

However the most important, and that which we as physicians have most frequently to deal with, is the respiratory function, which is divided into four parts; the first is, on account of the great vascular supply so near the surface, the inspired air is warmed before reaching the lungs; second, the air is purified by the arrest of all irritating particles which cling to the mucus covering the membrane; next, by moistening the inspired air; and last and not a minor function, the nose is the sounding-board or the resonance cavity of the voice.

Experiments have shown that air entering one nostril at a temperature of 46°, coming out of the other nostril, without entering the lungs, has a temperature of 86°, thereby showing the warming power of the cavities of the nose.

ILLUMINATION.

To look over the literature on diseases of the nose to-day, one would be led to think that this had been made a special study for many years, yet it was only in 1854 that Emanuel Garcia wrote of examining the nasal cavities by the aid of the sun-light and reflection, but practicing in London, as he did, fogs and darkness came so much oftener than sun-light that he abandoned this, and tried the oxy-hydrogen and electric lights, but these at that day, were crude and unsatisfactory.

Fig. 1.

Czermak of Pesth, a few years later, however, introduced laryngoscopy and rhinoscopy to the medical profession, using for his work the light of the student lamp. To-day, however, gas is most generally used in the form of the Argand burner. More brilliant still is the Argand burner covered with the MacKenzie condenser. (See Fig. 1.) However, doing work in a town where there is no gas, the student lamp, with the condenser, gives perfect satisfaction.

EXAMINATION.

Examining the nose anteriorly is called anterior rhinoscopy. Examining the posterior nares is called posterior rhinoscopy. The patient is seated in a darkened room with his back to the light; the light, that of an Argand burner or student lamp, being placed about one foot to his right and one foot behind and on a level

Fig. 2.

with his head. Now the patient being in position and the examiner in front of him, by means of a concave head mirror (see Fig. 2) a brilliant reflected light is thrown

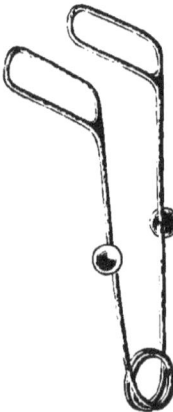

Fig. 3.

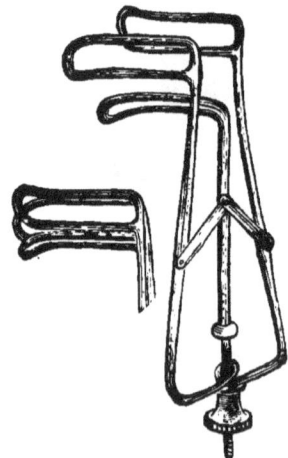

Fig. 4.

into the nostrils. To gain a good view, we dilate the nostril in the vertical direction by means of a nasal speculum. Nasal specula come in various forms and sizes; a simple two-valve is very good (see Fig. 3), however, in operations and different treatments, some physicians prefer a three-bladed speculum. (See Fig. 4.) By thus dilating the nostril and throwing in the reflected light, the nasal fossæ, the two turbinates, the septum and the floor of the nose come plainly into view; throwing the head backward by careful examination, one can see the roof and the superior turbinated bone.

POSTERIOR RHINOSCOPY.

The instruments necessary for examining the post-nasal space are a tongue-depressor, a post-nasal mirror; or, as it is more properly called, a rhinoscope, and some means to control the soft palate. There are many kinds of tongue depressors; the ordinary folding one of steel, nickel-plated and easily kept clean, is probably the

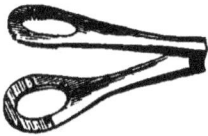

Fig. 5.

cheapest and best. (See Fig. 5.) There are, however, tongue depressors made of hard rubber, of glass, and of heavy wire, bent in various forms, from which one may select. (See Fig. 6.)

In introducing the tongue depressor, the examination must be made lightly and not too firm pressure made at the base of the tongue, otherwise irritation will be produced, thus causing retching and gagging in many cases.

The rhinoscope is a small plain mirror on the end of a long wire with a handle. The mirrors come in several sizes, but the size used in this examination is the No. 1 mirror of small diameter. (See Fig. 7.) In making a

Fig. 6.   Fig. 7.

thorough examination of the vault of the pharynx and
posterior nares, the chief difficulty presenting is the inter-
.erence of the soft palate. To overcome this there are
numerous palate retractors, but I prefer a round rubber
cord, such as is used for a ligature in minor amputations.

Seat the patient in position, pass the rubber cord along
the floor of the nose until it passes down into the throat ;

now opening the mouth, the cord can be seen hanging in the throat in full view. With a pair of forceps grasp the cord and draw it forward out through the mouth.

We now have the complete circle, and making gentle traction we notice that the soft palate is drawn forward, when the two ends of the cord are tied in a knot on the lip.

This greatly augments the working space in the throat, produces no retching, and the examination, most difficult before, is now rendered very simple.

The mirror is then heated by holding the glass surface over the light for a few seconds, the tongue is depressed, the rhinoscope introduced into the throat at the proper angle and the light from the head mirror is thrown onto the rhinoscopic mirror, in which the image is seen ; by changing the position of the rhinoscope different views can be seen ; by this means we can, with certainty, examine the vault of the pharynx, the posterior portions of the turbinates and the orifices of the eustachian tubes.

DISEASES OF THE ANTERIOR NASAL CAVITIES.

ACUTE RHINITIS.

Acute rhinitis is known also by other names, as acute coryza, acute nasal catarrh, acute nasal blennorrhea, snuffles, and cold in the head.

SYMPTOMS.

As the name implies, rhinitis signifies inflammation of the nasal mucous membrane ; acute signifies a short space of time, or an inflammation which is rapid in its course. It comes on by a sense of dryness and fullness in the

nostrils, intense itching, and sometimes sneezing. The
fullness is due to the engorgement of the blood vessels
and the general hyperæmia of the parts. The disease
presents itself in different grades of severity. The skin
is hot and dry and the secretions are more or less
arrested, and in more delicate persons there is at times
considerable fever. Within two or three days, or sometimes longer, comes on the *second stage* or the stage of
dilatation of the blood vessels; there is an abundant
watery discharge from the nostrils, scalding the nose and
the lips, and often there is increased lachrymation, due
to irritation of the eyes.

Generally in this stage, and sometimes even in the
first stage, there is intense frontal headache at times
almost unbearable, due to implication of the frontal
sinus.

This inflammation of the frontal sinus can be diagnosed by the frontal headache, the increased temperature, and by eliciting tenderness and pain on pressure
over this region.

The voice assumes a nasal twang so noticeable in this
disease, and at times there comes on an impairment of
hearing, due to the close relation between the nose and
the middle ear. The disease progresses and finishes by
a lessening of the fever, the restoration of the secretions
and the nasal discharge becoming heavy, thick, yellowish
in color, consisting of mucus mixed with pus.

ETIOLOGY.

The causes of this disease are many and varied, the
commonest being too great heat or overexercise, followed
by rest or exposure; also the inhalation of irritating
dust or vapors; also of hot dry air, for instance where a
house is too hot from a hot-air furnace, etc.

In scarlet fever and measles acute rhinitis is always present as a constant complication; or better, a diagnostic point in these diseases.

There is no doubt, also, but that a scrofulous taint in the system predisposes one to this disease.

PROGNOSIS.

The prognosis is always good as far as life is concerned; also the prognosis is good for a cure, but it is well to remember the complications of the various sinuses, the relapses and new colds contracted, and especially to remember that this disease, in repeated attacks, in nine cases out of ten, forms the nidus or starting point for the various forms of catarrh.

TREATMENT.

When one thinks of treating a disease he generally thinks whether the disease can be aborted. This, however, seldom comes under the charge of the physician, as the patient himself generally has some panacea at home. Among these might be mentioned hot drinks, hot lemonade, the coal-tar products, and most frequently quinine.

There is no doubt but that the bad results of an exposure may be aborted by taking a hot foot-bath and ten grains of quinine and immediately retiring, warmly wrapped in blankets; however, the habitual use of quinine in this country is a most pernicious habit, and we have much better agents than quinine, and quinine and whisky, and the combination of rock candy, whisky and glycerine. If the patient is determined to have quinine there is one *contra-indication* for its use, viz: When there is irritation of the nervous system, associated with a harsh, dry skin. If called early in the case one of the best means to hasten a cure is to prescribe a hot mustard foot-bath, take a fair sized Dover's powder and then retire and arise late the following morning. Those who do not like Dover's powder may use aconite and asclepias. In this stage, if there is fever and a full bounding pulse, there is no agent better than veratrum. Determination of blood to the head, accompanied by irritation of the nervous system and a watery discharge from the nose, there is nothing better than gelseminum and rhus. In implication of the frontal sinus with the extreme frontal headache, bromide of ammonium will cure; in case the patient is obliged to go about his business, a *protective* is necessary to the lining of the nose; let him use a spray of liquid albolene. When there is considerable pain and *stenosis* (occlusion or stopping up of the nose) he may spray the nose every three hours with a two per cent. solution of cocaine.

In the *stage of resolution*, where there is the great quantity of muco-purulent discharge, he may use a spray or nose wash, or sniff from the palm of his hand the following:

R—Bicarbonate soda.
Biborate soda.
Common salt, *aa.* 3 j.
Water, O j. M.

After cleansing the nose with this wash, always spray the nose with albolene before going out. Especially warn the patient that this is the time when he is most liable to renew his cold, and he must therefore guard himself carefully.

In case any bad odor manifests itself at this time, one of the ingredients in the above wash may be changed to one of the following: Either potassium permanganate, 10 grains to the pint, or chlorate of potassium, 1 drachm to the pint, or carbolic acid, 10 drops to the pint. While this particular disease gets well of itself, yet it is so aggravating that patients come for treatment, and the above mentioned local and internal treatment is not only soothing, but brings the disease to an early termination.

SIMPLE CHRONIC RHINITIS,

Coryza, chronic nasal catarrh, purulent catarrh, etc. This simple form of chronic catarrh is caused in the main by repeated attacks of the acute rhinitis. It also comes from continued inhalation of irritants, as in a person employed in a cooperage establishment, or seed house, and from the irritating effect of chemicals and tobacco smoke.

PATHOLOGY.

The continued irritation produces vascular engorgement, which, followed in its turn by dilatation and relaxation of the vessels, leads to exudation and loss of certain blood elements.

Most noticeable is the escape of white corpuscles, some of which lodge under the epithelial layer of the mucous membrane, and, organizing somewhat, form an induration; others break down and form pus cells, and mixing with the increased mucous or serous secretion, we have the thick, heavy, characteristic catarrhal discharge.

SYMPTOMS.

The patient complains that he has a chronic cold in the head, with a large amount of muco-purulent discharge, not only anteriorly, but also posteriorly, and heavy, yellow or green or blood-tinged chunks of purulency.

He has a nasal twang to his voice, complains of a more or less stuffed-up feeling in the head (stenosis), and he often awakes in the morning with a dull, heavy, frontal

headache. There may be so much irritation anteriorly
that he picks his nose with his fingers, causing bleeding
and ulceration, which sometimes extends and causes
perforation of the septum. Implication of the frontal
sinus or of the eustachian tube is not uncommon in the
acute exacerbations of this chronic form. While there
may be more or less swelling of the turbinated tissue,
still, on the application of cocaine, the membrane re-
tracts to its normal size, showing that it is not a case of
hypertrophic rhinitis.

PROGNOSIS.

The disease, left to itself, remains about the same, or
ends in hypertrophic rhinitis. Carefully treated, however, the prognosis is favorable, providing you can
change the patient's occupation away from irritants.
One could hardly expect to retain the irritating cause
and at the same time produce a cure.

TREATMENT.

In treating disease of a catarrhal nature, the first
point is to put the patient on the hygienic treatment
for catarrhal disease in general, especially if it is a bad
case, or where the patient is extremely sensitive to exposure—that is, he takes cold easily. This treatment
means that he must wear substantial woolen underwear
the whole year. It is in catarrhal troubles that the
Jaeger underwear has its sale. There are other makes
as good, but I just mention this one. Next, the patient
must, each morning, sponge the neck, arms and chest,
with cold water or salt water, followed by brisk friction
with a coarse towel. He must not wet the hair; must
have light or medium diet, and pay special attention to
the regular action of the bowels. Astringents in the

nose are not well borne. The best effect comes from a cleansing, alkaline and antiseptic solution.

One of the best preparations on the market is "Seiler's alkaline and antiseptic tablets." Dissolve one tablet in two ounces of water and use as spray. A cleansing and alkaline spray solution can be made by dissolving borax, bicarbonate soda, and common salt, each one drachm in a pint of water, to which, if you choose, you may add a few drops of carbolic acid or a small quantity of asepsin or salicylic acid.

Spray and cleanse the nose thoroughly two or three times daily with the above, *after which* use the following oil spray :

R—Menthol, camphor gum, *aa*. grs. x. Rub together until a liquid results, then add liquid albolene, q. s. ℥ j. M. (See Fig. 8.)

Fig. 8.

I know of no oil or petroleum that can replace the albolene, and I know of no combination that has given me satisfaction equal to this one. In fact I have seen catarrhal cases get well on this alone.

In case the patient has an acute exacerbation, with

pain, swelling, and stenosis, this may be temporarily overcome by a spray of cocaine, but in chronic cases it is to be avoided.

In case there is much hyperæsthesia and vascular excitement, a good sedative spray having somewhat anæsthetic properties is the following :

    ℞—Potassium bromide, grs. xv.

       Water, ℨj. M.

Sig.—Use as a spray every few hours.

The *carbonate of calcium*, the second trituration is given in about 5-grain doses when the mucous membrane of the nose and throat is red, and especially sensitive to dust, irritants and to all forms of local applications.

The *phosphate of calcium*, second trituration is indicated in anæmic patients, especially where there are adenoid vegetations in the vault of the pharynx, or a relaxed throat or mucous polypi in the nose.

The *sulphide of calcium*, second trituration is indicated where there is a heavy purulent discharge : in catarrh associated with eruptions of the skin, and where the lymphatic glands of the neck are enlarged and hard.

Penthorum sedoides is indicated where there is a sense of fullness of the mucous membranes with abundant secretion; spongy gums; fullness of the throat and ears.

Pulsatilla is given where the discharge is thick and yellow and does not excoriate; the usual systemic indications also act as an aid in selecting this remedy.

HYPERTROPHIC RHINITIS.

This disease in its various forms is probably more frequently met with than any other nasal disease, unless, perhaps, it be the simple chronic rhinitis described and treated in the previous chapter.

This is the disease which causes so much occlusion of the nasal passages, called *stenosis*, such a stuffed up feeling, so often producing habitual mouth breathing, a pernicious habit, which, of itself, causes a long line of evils.

ETIOLOGY.

To describe the course of a bad case, the disease is brought on by a susceptibility to cold, to irritants of all kinds, dust, tobacco smoke, coal soot, etc.; chronic rhinitis also merges into this disease; it also results from irritating snuffs, lotions and the forcible use of the nasal douche.

This nasal mucous membrane being over-sensitive and subjected to the above irritants, there is the increased flow of blood to the part, the corpora cavernosa become engorged, there is dilatation and relaxation of the vessels following, consequently more or less exudation and new tissue elements formed, which, with time and repeated irritation, cause great enlargement, thickening and swelling of all the mucous membrane, *particularly that covering the turbinates.*

In this disease especially, does this sub-mucous or vascular layer play a great part, and it has been well described as the corpora cavernosa of the nose, under the influence of the sympathetic.

The coats of the venous sinuses become indurated, new tissue is formed round about, so that the vessels can not empty themselves, but remain distended, and this condition, together with the new tissue and the general enlargement, encroach upon the breathing space, this being the real cause of the many annoying subjective symptoms.

SYMPTOMS.

The symptoms vary much in different cases, but, as a rule, the patient complains of a " stuffed up " feeling in the nose; he has not room enough, space enough to breathe through it.

The discharge in some cases amounts only to a hyper-secretion, at times less than normal. However, in a case well advanced, there is liable to be ulceration in some form, due to parts touching each other, or from the presence of polypi, which are not infrequent.

The patient frequently complains of morning head-aches and sometimes of asthenopia. He complains particularly that on lying down, on the right side, the right nostril is occluded, and *vice versa*, to remedy which he breathes through his mouth, and each morning goes through the usual routine of hawking, spitting and gagging in order to dislodge the secretion and debris that have formed in the post-nasal space.

MOUTH BREATHING.

Mouth breathing as a consequence of hypertrophic catarrh, is quite a common occurrence, which in its turn has a train of disagreeable results. The most common result is follicular pharyngitis; the follicles become enlarged, red and painful, and especially in the mornings on awakening, there is a *dryness* and irritation which is particularly painful.

In confirmed mouth breathing, there comes on a stupid and silly facial expression, a nasal twang to the voice, and not infrequently a catarrhal laryngitis associated with cough that in many cases forms the irritation, the starting point, the nidus for phthisis.

TREATMENT.

The treatment is both medical and surgical, and in treating catarrhal troubles I have learned that it is far better to treat a patient three times a week for three weeks, than once a week for nine weeks. By this I mean, treat a case *thoroughly*, conscientiously and carefully at short intervals and get through, rather than at long intervals, thereby covering a long space of time and often not getting a cure; also for the reason that patients with catarrh are difficult to hold, and unless they see that you are actively treating them, they will stop treatment.

It is an *axiom* that patients treating for catarrh should come in every other day.

The patient with hypertrophic nasal catarrh has the same disease as treated in the previous chapter, chronic rhinitis, and should have precisely the same treatment— cleanliness, sprays of special nature, hygienic and systemic treatment—but in addition he has the hypertrophy which requires surgical means.

The surgical means referred to is generally cauterization. By that I do *not mean searing* everything with which you come in contact, producing thereby an atrophy and dryness for which the patient will curse you the rest of his days, but I do mean making an antero-posterior streak or scar throughout the middle of the inferior turbinate, thereby practically losing *no* mucous membrane, and by this line or cauterization to cause a cicatrix which lessens the hypertrophy and shrinks the enlargement, thereby overcoming the stenosis and giving a free nasal passage.

The first agent I shall mention is the galvano-cautery. Use as follows: Place a good sized pledget of cotton

8

saturated with a four per cent. solution of cocaine, in the nostril to be operated on, and at the end of six or eight minutes complete local anæsthesia will supervene and the working space will be greatly enlarged.

Nostril well dilated and patient's head steady, introduce the cautery knife cold and carry it to the back part of the turbinate, then closing the current, the hot knife is brought forward to the anterior part and out again. This makes a cut as deep as you choose with no pain nor bleeding, providing the knife is at *cherry red* heat, *not* black heat nor white heat. (See Fig. 9 and 10.)

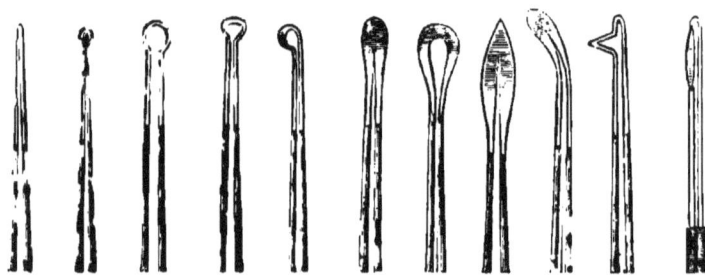

Fig. 9.

Fig. 10.

The after treatment consists in spraying the nostril
twice a day with some alkaline and antiseptic solution
(see Fig. 11), and each day when they come in the office,
see that there is not too much ulceration, in which case
you might swab the part with a solution of permanganate of potassium or insufflate aristol. (See Fig. 12.)

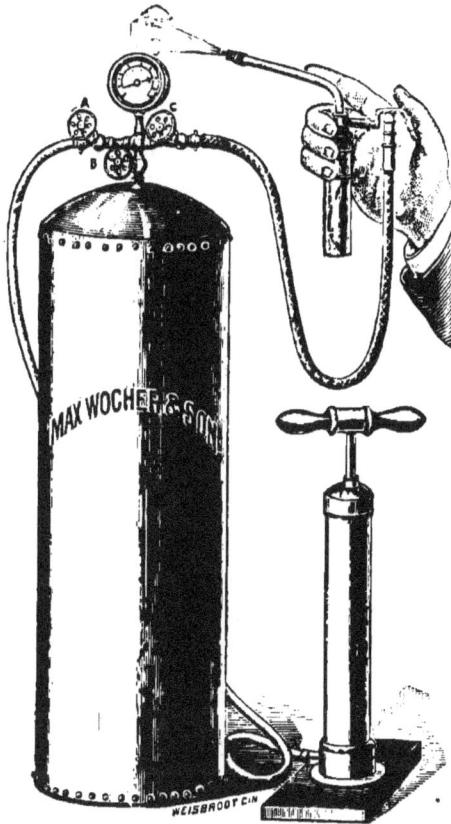

Fig. 11.

This careful after treatment is of the greatest importance, insuring a nice, clean cicatrix and a good result.
In case you have no cautery, the same result may be

reached by using chromic acid. The technique, cocain-izing, and formation of scar are exactly the same as above.

Chromic acid comes in the form of red brown crystals, and is applied by means of a silver probe; that probe which comes in the pocket surgical case is very good.

Heat the end of the probe quite hot, then touching the chromic acid, some crystals adhere to it, when again holding the probe over the heat, the crystals melt and a round bead is fused on the end. Now dry the turbinate with cotton and use this fused chromic acid, making the scar in the manner above described.

Immediately on finishing, neutralize and destroy all remaining chromic acid in the nose by swabbing the parts with a sat-urated solution of bicarbonate of soda.

The after treatment is the same, and scab comes off several days after, and nothing is noticed but the lineal scar with its consequent shrinking.

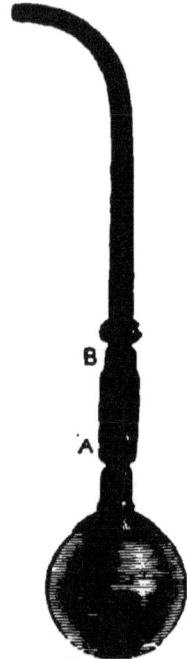

Fig. 12.

This procedure may be repeated at intervals of two or three weeks until the hypertrophic tissue is sufficiently reduced, and the nasal passages become again patulous. In cases of a fair constitution and where they will follow directions and treatment, the prognosis is very good.

INTERNAL TREATMENT.

The internal treatment of hypertrophic rhinitis is rather limited, the entire treatment being almost wholly local. When the mucous membrane is very red and irritable, and especially where there are frequent attacks of epistaxis carbo. veg. given internally is a most excellent agent.

The diet in all cases should be light and good, special attention should be paid to the regular action of the bowels, and the patient should have an occupation free from irritating dust and exposure.

ATROPHIC RHINITIS.

Atrophic rhinitis is known also by the names dry catarrh, atrophic catarrh, ozena, etc., and of all the forms of catarrh is most to be dreaded.

This unfortunate condition seldom begins after middle life, but even contracted before, may last a lifetime.

This is the disease which people generally think of when speaking of catarrh; although there are so many varieties, their mind immediately flies to the thought of a fallen or sunken nose, the fearful stench and quantities of purulent secretion.

The origin of this disgusting disease is thought by many to be due, in an indirect way, to a strumous or syphilitic diathesis; or, in other words, that it is the manifestation of some constitutional taint.

I do not wish to be understood as saying that *every* case is constitutional, but I do know that nine out of every ten cases require thorough constitutional treatment.

SYMPTOMS.

The symptoms, while few, are so constant and severe that it is well there are no more.

Dryness is always constant; the patient seemingly would undergo any treatment could he but be relieved of the sense of dryness and consequent irritation.

The principal symptom complained of is the abundant secretion from the nostrils, purulent or of a yellow or greenish hue, which comes away in the form of lumps or large scabs.

Accompanying this secretion is the stench, disgusting and loathsome, which ostrasizes him from the society of his friends.

This odor is peculiar and characteristic, and when once smelled.is always known thereafter.

This retrograde change, with its constant secretion, almost gangrenous in character, at times sets up violent headaches, accompanied with fever.

There is no doubt that this disease has a bad influence on the general constitution of the patient, and particularly on the stomach.

PATHOLOGY.

The most noticeable change is the atrophic condition throughout the nasal fossæ.

The membranes present a light colored appearance, and the loss of the healthy capillary circulation is apparent.

There is consequently a loss or a breaking down of the mucous and serous glands, thereby losing the natural lubricating fluid, producing the dryness complained of and leaving the sensitive mucous membrane entirely exposed to the action of all irritants, to say nothing of the lost nasal function of purifying and moistening the inhaled air.

On account of this poor capillary circulation and dryness there ensues a superficial dessication, or, as perhaps better described, ulceration takes place in different places, in patches here and there. There is consequently the formation of large scabs, consisting of mucus, and broken down tissue of a purulent nature, accompanied by the characteristic stench.

In more unfortunate cases, the ulceration becomes more or less phlegmatic in character, and the bones soon become affected, and necrosis of the bones and perforation of the septum, etc., are not uncommon.

In proportion to the general atrophy of the parts, there is the consequent shrinking, and the breathing space in the nose is proportionately increased.

PROGNOSIS.

The prognosis should in all cases be governed by the severity of the disease and the constitution and general condition of each patient.

There are cases which can be cured; there are many which can not, but you can improve all.

In alleviating the worst symptoms, overcoming the foul breath and headaches and partly establishing normal secretion, I am satisfied that this patient will get more benefit from his treatment and be better satisfied, than in the treatment of less severe diseases.

TREATMENT.

In the treatment of this affection the first step is get rid of the crusts, scabs and dry formations which are found in such great quantities, or, in other words, cleanliness is first.

Sprays of various kinds would only medicate the surface of the crusts; it takes something more powerful to remove them; this agent is the *douche*. (See Fig. 13.)

DIRECTIONS.

A common rubber bag, such as comes in various forms, with a hard rubber post-nasal tube, forms the most convenient article.

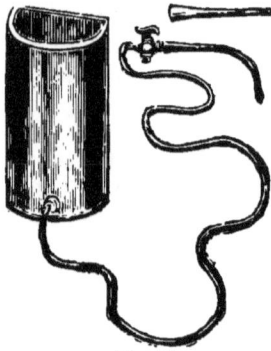

Fig. 13.

1. The solution used in the douche must be *alkaline* in reaction, to loosen and remove the scabs.

2. The temperature of the liquid must be a little more than lukewarm, about 98° to 100°.

3. The douche must not be used with too much force; hang the bag just a trifle higher than the head.

4. Do not blow the nose violently, nor go out into the air soon after using the douche unless previously spraying the nose with some oil, preferably albolene.

The patient should use the douche at least twice a day, taking care about the above directions, in which case no middle ear trouble will follow.

A very good and economical prescription for a douche is the following :

R—Powdered borax.
      "   bicarb soda.
      "   common salt, *aa.* ℨj.

Give the patient a number of these powders, with directions to take one powder, dissolve it in a glass of hot water, then pour same into the douche, and add a pint or a pint and a half of water, and see that the temperature of the present solution is about 98°; it is not necessary that it be *exactly* 98°, but it must be *neither* cool nor hot.

In cases where the extreme ozena or stench is the prominent symptom, a good combination is to take the above three agents and add either permanganate of potassium 10 grains, or the sulpho-carbolate of soda ℨj.

However, for the stimulation of the vessels, and to encourage the glandular structure, nothing is better than ammonium chloride, as follows:

R—Borax ℨ ii.
Ammonium chloride ℨ ii.
Potassium permanganate grs. xx.

Sig: Dissolve in one quart water, temperature 98°, use as douche night and morning.

Many patients find relief from headache and perfect stimulation to the nasal mucous membrane by using the following as a *spray*:

R—Menthol grs. x.
Water ℥ j. M.

A good office treatment consists of spraying and cleansing the mucous membrane with some alkaline and antiseptic solution, then touching the ulcerated areas with a ten grain to the ounce solution of nitrate of silver on a cotton carrier, after which spray the nose thoroughly with albolene and insufflate aristol.

The aristol being caught by the oil, will remain in contact with the membrane for a long time.

Many cases, especially those taking cold easily, find relief in a spray containing several drops of each of the following: Oil of tar, thymol and eucalyptol to the ounce of albolene. Use three times a day.

A case can not be cured without special care to the general health, and here tonics in general and specific medication in particular, play a great part.

There is generally an indication for sulphite of soda, which works wonders. I have even used it locally with good results.

In so far as there is so much to gain, such unpleasant symptoms to overcome, every case is treated with benefit and satisfaction.

SYPHILITIC RHINITIS.

ETIOLOGY.

As indicated by its name, this disease of the nose is due to constitutional infection. It frequently occurs in the secondary stage, also in the tertiary stage after the lapse of fifteen or twenty years. It also occurs as a result of hereditary syphilis.

SYMPTOMS.

Upon examination of the nasal cavities, ulcerative patches are noticed, erosions and grey spots which have a tendency to spread and become phagedenic in character. The term phagedenic signifies active, rapid, eating away. The ulceration does not stop here, however, but often attacks the deeper structures, attacking first the periostium and then the bones. The disease attacks the turbinated bones and soft tissues, but most frequently the septum; first, the anterior cartilage, causing the tip of the nose to fall, and then the bones, in which case the foundation being destroyed the bridge of the nose is flattened and the patient is disfigured.

Scrofulus rhinitis is on the same order, but milder, bearing the same relation to it throughout, as does scrofula to syphilis.

PROGNOSIS.

A patient whose tissues begin to be attacked by an ulcerative process is indeed to be pitied, and in this case must necessarily be in a very low state of vitality, consequently the prognosis must be guarded.

All cases of syphilitic disease of the nose do not end in this unfortunate manner, for where the patient has any system to work on, local measures and internal treatment will check the disease.

TREATMENT.

The ulcerative process and the constant dessication of the scabs produce an odor which can not be mistaken; it is similar to that of atrophic rhinitis, but worse.

The common sense course of treatment is local cleanliness and internal anti-syphilitic treatment.

By cleanliness I do not mean to insufflate antiseptic powders, nor to use sprays; the only radical means is to *wash out* the scabs dead material and debris, and produce a healthy nose by their absence rather than by covering them over with another smell. To accomplish this the douche is the only means, and use it as directed under atrophic rhinitis.

A prescription that has the proper qualities and will also prove a disinfectant, is the following ·

> ℞—Borax.
> Common salt.
> Sulpho-carbolate of soda, *aa.* 3j.
> Water, Oj. M.

Use the above quantity in a douche at least twice or three times daily.

As an extra antiseptic, the nose may be sprayed with albolene and then aristol insufflated, the aristol clinging to the oil and thereby remaining in contact with the membrane for several hours.

The necrosed bone had better be scraped with a curette, or if loose taken out, and the ulcerative patches cleaned and touched with stick nitrate of silver or nitric acid.

Little can be accomplished without internal medication, then the physician may choose from his list of alteratives, iodide of potassium, stallingia, iris, Donovan's solution, arsenic, the vegetable alteratives, giving them according to indication.

Strange to say, many patients do get well and have no recurrence of the trouble.

NASAL POLYPUS.

Among the tumors or growths occurring in the nasal cavities, the most frequent is probably polypus.

Polypi are divided into two varieties, mucus polypi and fibrous polypi, or technically speaking, *myxoma* and *fibroma*.

PATHOLOGY.

Polypi have their origin in the mucous membrane and most frequently spring from the lower surface of the middle or inferior turbinated bones, sometimes from the superior turbinate and from the various sinuses.

The attachment is small and pedunculated and they hang downwards in the shape of a pear, the large ones assume the shape of the cavities to which they are confined.

MUCOUS POLYPI grow by an increase of the tissue elements both of the sub-mucous and the epithelial layer; they are soft, gelatinous in character, and of a pearly white color.

FIBROUS POLYPI most frequently grow from the roof of the nose and the posterior nasal space, and are much more serious than the mucous.

They arise from the periostium and sometimes from

the bone itself, are fibrous and tough in character, and are freely supplied with blood vessels.

SYMPTOMS.

The symptoms of polypi are very few, many patients have no suspicion of their presence until an examination is made by the surgeon.

They may notice that in damp weather there is an increased sense of fullness in the nose. This is due to the hygroscopic nature of the mucous polypus.

The patient may notice that his nose is occluded and on account of the closure suspect some growth.

From pressure, on account of its size, a polypus sometimes produces headache, sneezing, epistaxis, and not unfrequently reflex coughs, asthma and facial neuralgia.

The two varieties are easily diagnosed; the *mucous* polypi are soft, pearly white, look exactly like an oyster and change in damp weather; the *fibrous* are hard and red in color.

TREATMENT.

In so far as polypi, in almost every instance, owe their origin to some catarrhal trouble, the first treatment must be directed to this, that is alkaline and antiseptic treatment.

The only radical treatment is operative, of which there are several ways. The first and that most frequently employed is *torsion*.

The side to be operated upon is thoroughly sprayed with a four per cent. solution of cocaine, after which an interval of a few minutes is allowed for its absorbtion.

Using strong reflective light the polypus is grasped with the polypus forceps, and by a twisting motion is withdrawn.

The other and best means is that of the wire snare. (See Fig. 14.) The polypus is snared and the wire is pushed as close to its base as possible, when the loop is *slowly* closed.

This means causes almost no pain or hemorrhage, while by torsion there is always pain and considerable hemorrhage, which, however, is never serious. This treatment refers to mucous polypi.

Fibrous polypi, however, are more formidable and really come under the head of major surgery.

When large and occurring in the roof of the nose, the operation must be done under chloroform, the upper lip and nose itself are dissected from their bony attachments and turned upward on the forehead, when the whole cavity is entirely exposed.

On account of the great vascularity of the fibrous polypus, and its location, the *galvanic snare* is the best means of removal, as hemorrhage is less likely to occur. This operation offers the advantage of leaving no scar, compared with other operations of bony detachment of the nose externally.

When the fibroma is large and hangs down into the pharynx from the post-nasal space, the soft palate must be split and the tumor removed by way of the mouth.

It is well not to lay too much stress on a favorable prognosis, as recurrence after thorough operation is not uncommon.

MAX WOCHER & SON, CIN., O.

Fig. 14.

ECCHONDROMA.

Ecchondroma is a cartilage tumor or growth springing from the septal cartilage.

On account of the anterior location they are easily diagnosed and also easily removed.

While it is true they are pathological, yet they seldom do any harm except in singers and public speakers, in whom a change in the voice may be noticed.

The *treatment* is surgical, cocainizing the part and either removing the growth with a small nasal saw, or by the use of a cartilage knife. (See Fig. 15.)

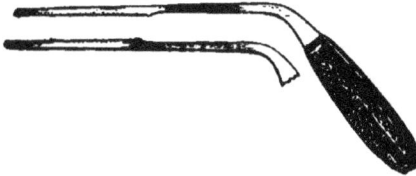

Fig. 15.

The operation is always successful and the improvement in the voice is sometimes considerable.

EXOSTOSIS.

Exostosis, the common name of which is *a spur*, is a spur or spicula or bony growth springing outward from the septum.

When small they are of no importance, however they sometimes act as a source of irritation to the eyes, producing asthenopia; they are said in some cases to produce reflex asthma, and in singers by projecting into the nasal cavities they injure the tone of the voice according to their size.

These are taken off by the nasal saw, and by some
operators, by means of the dental engine and special
burrs.

Under cocaine there is scarcely any pain or hemorrhage.

DEVIATION OF THE SEPTUM.

Anatomists in general describe the nasal septum as
bending somewhat to the left, but whether right or left,
it is the exception rather than the rule to find the septum
straight.

The deviation or deflection can only be considered
pathological when it bends so far to one side that it
encroaches so much upon the space as to produce more or
less *stenosis*.

When one side is practically occluded by the deviating
septum, then an operation is indicated.

The operations for straightening a deviated septum are
so numerous that I shall here only mention the one most
frequently used and the most simple.

After thoroughly cocainizing the nose, by means of a
strong knife or nasal punch (see Fig. 16), the septum is
split in the *axis* of the curve or deviation.

By pressure, from the occluded side, the septum is now

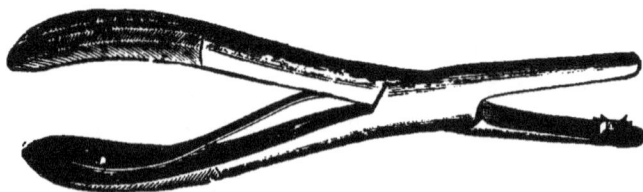

Fig. 16.

forced into a straight position (see Fig. 17), when the cut surfaces over-ride each other.

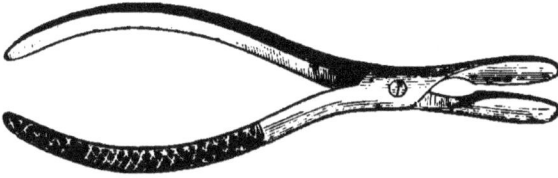

Fig. 17.

The side which was occluded is now packed tightly with oakum or absorbent cotton made antiseptic, in order to hold the septum in its new position.

This plug must be removed by the surgeon each day and the cut surfaces swabed with some antiseptic solution.

The operation and after treatment must be carefully carried out or the results will be negative.

EPISTAXIS.

Nose-bleed is a rather common occurrence and comes in a great many cases from blows and injuries of different kinds, however, following bad injuries of the nose the patient is subject to frequent attacks for quite a number of years; but even in these cases and others of nose-bleed the cause is most frequently congestion.

Among exciting causes may be mentioned blows, falls, blowing the nose, sneezing, picking the nose, bleeders, vicarious menstruation, old age, etc., etc.

The hemorrhage is, as a rule, one-sided and either anterior or posterior, which can be diagnosed by noticing whether the blood flows out through the nostrils, or back into the throat when sitting in an upright position.

9

TREATMENT.

Among minor measures may be mentioned first, pressure of the bleeding side against the septum, holding the arms above the head, a hot foot bath and sniffing ice water. In a case of epistaxis these may be tried in their turn, but if not successful, one may use powdered tannin and alum by insufflation.

As a last resource in severe cases *plugging* must be resorted to. In lieu of a Bellocq's canula a soft rubber catheter may be used, the technique is as follows: pass the catheter backward through the nose, until it hangs down in the throat, the patient then opening his mouth, the catheter is seen and drawn forward with forceps, to this end of the catheter the string, to which the plug is attached, is now tied; the catheter is then withdrawn the way it entered, the string following, until the plug rests in the posterior nares, and being too large to pass through thoroughly plugs the same; now plugging the anterior nares the hemorrhage is within bounds and soon ceases.

The posterior plug should also have an attached string leading outward through the mouth by which it may be withdrawn; these dressings should not be allowed to remain over twenty-four hours, on account of decomposition and consequent trouble.

HAY-FEVER.

The name in use to-day for hay-fever, hay-asthma, rose-cold and pollen catarrh is *periodical hyperæsthetic rhinitis*.

This is an acute rhinitis with hyperæsthesia of the mucous membrane coming on at a regular time each year, sometimes in June but generally about the twelfth or fourteenth of August.

The individual is susceptible to all irritating particles in the atmosphere, there is a burning watery discharge, a stuffed up feeling in the head and extreme redness and irritation which extends sometimes to the eyes, ears and bronchi, even at times producing a typical asthma.

TREATMENT.

In so far as catarrh underlies the majority of cases of hay-fever, the expected attack is best cut short by attending to this a few weeks before.

Internally one of the best agents is *napthalin*, the second trituration, given before and during the attack; it is sometimes a preventative.

Arsenic is given where the discharge is profuse, watery and excoriating.

Rhus tox is given where the patient complains particularly of the burning.

Aconite is indicated in the first stages of the disease.

The first local agent I would recommend is *camphor-menthol*, ten grains each to the ounce of albolene, used as a spray; this has given me the best satisfaction.

Peroxide of hydrogen is next to be tried, one part to three of water. If there is any ulceration the patches should be cleaned and cauterized.

Cocaine is not prescribed as a chronic course of treatment, but may be used as a spray, once or twice a day, when the disease is at its worst.

It must not be forgotten that this is a most painful and annoying disease, and that results are frequently negative, but that in the end, the patient may be sent away from his work on a vacation or sea voyage, and be much benefited.

DISEASES OF THE POSTERIOR NASAL CAVITIES.

ACUTE PHARYNGITIS.

Acute pharyngitis is generally what is called sore throat, or is present at the same time.

On account of the continuity of tissue, a patient seldom has an acute pharyngitis alone, but has associated with it trouble of the nose, the larynx or the tonsils.

The disease soon gets well of itself, but in bad cases, we are sometimes called, when aconite, veratrum, phytolacca or belladonna have their indications.

POST-NASAL CATARRH.

There is, of course, a disease called chronic pharyngitis, but as post-nasal catarrh is so prevalent, and of so much importance, and so similar to it, I shall speak of it instead.

Post-nasal catarrh is also known by the name of chronic catarrh of the naso-pharynx, follicular disease of the naso-pharyngeal space, American catarrh, etc., etc.

ETIOLOGY.

The disease follows repeated attacks of acute pharyngitis or sore throat; this is the most frequent cause.

It follows various forms of catarrh of the nose and is also produced by mechanical obstruction of the anterior nares; that is, the secretion that should naturally pass forward and out, flows backward and is a constant source of irritation. It frequently follows scarlet fever, measles and diphtheria.

PATHOLOGY.

All around the orifices of the eustachian tubes, on the soft palate and where the septum joins the palate posteriorly, we find the conglomerate glands which constantly throw out a viscid mucus.

On the posterior wall of the pharynx we find the follicular glands, large and numerous, and it is owing to the chronic or subacute inflammation of these glands, that there is so much secretion and *dropping* into the throat and why a cure is effected so slowly, because an impression on glandular structure is most difficult to make.

SYMPTOMS.

The chief symptom complained of is a constant dropping in the throat, so that at all times the patient has a desire to hawk and clear the throat. It is wonderful the amount of secretion that can come from so small a space. There is a huskiness of the voice peculiar to this disease. In some cases the tissues of the throat are relaxed and flabby, while in others they are hyperæmic and irritable.

In almost every case there is more or less dyspepsia; whether the catarrh produces the dyspepsia or the dyspepsia produces the catarrh, it is certain that the general health is affected and that one can not be cured without treating the other.

TREATMENT.

One of the easiest means to reach this region is the gargle, for which there are many prescriptions, each physician having one which he thinks is best, nevertheless the gargle must be alkaline, antiseptic and astringent.

In so far as this is generally associated with nasal catarrh, I generally prescribe camphor-menthol as a spray.

In regard to astringents, one must use a strong solution, if he would effect the glandular tissue in this region. One of the best means is to make a post-nasal application every other day of the following:

R—Nitrate of silver, grs. XL.
Water, 3 j

There is another prescription, however, that is used to a very great extent:

R—Iodine, grs. iii.
Iodide of potassium, grs. xii.
Glycerine, 3 j.

When the follicles in the throat are particularly large and prominent, they may be touched with nitrate of silver, one hundred grains to the ounce, or use the galvano-cautery needle, touching only three or four at a sitting.

In case there is a relaxed throat, the patient may gargle with fluid hydrastis, pinus canadensis or hamamelis, one part to six parts of water.

Internally, to overcome the dyspepsia and sweeten the breath, sulphite of soda is given.

Where the patient has catarrhal trouble associated with boils, or enlarged and indurated lymphatic glands, sulphide of calcium is given, second trituration.

Podophyllin, however, is considered to be almost a specific in chronic throat diseases where there is fullness of tissues, and the characteristic tongue.

DISEASES OF THE THROAT.

ACUTE TONSILITIS.

The common name of acute tonsilitis is quinsy. It generally effects children and young adults, and is rare after thirty and in late life.

ETIOLOGY.

Ordinary exposure is the common cause of quinsy; however, ænemia and scrofula play no small part.

SYMPTOMS.

One or both tonsils may be inflamed, swollen, red and very painful on deglutition. In some cases the disease is preceded by a chill and fever, the fever running quite high, and the secretions arrested. The attack lasts from three to ten days, and usually ends in resolution. The prognosis is good and the diagnosis is easy.

TREATMENT.

Internally aconite is the remedy in the first stages and in children.˙

Belladonna is given where the throat is red, dry, and congestion well marked.

Veratrum is given where the inflammation is sthenic in character and where there is fear of suppuration.

Phytolacca is given probably more than any other agent. The throat is sore, the tonsils enlarged perhaps the lymphatics of the neck, and there is a paleness of the mucous membrane.

When the inflammation and swelling become intense they may be controlled by painting the tonsils with veratrum. The inhalation of steam containing vinegar is excellent where there is a dryness and a choking sensation. Stillingia liniment, two drops on a lump of

sugar, and rubbed on the outside will in most cases relieve the tickling sensation that causes so much coughing. With the above named treatment the disease almost
always ends in resolution, only once in a long time will
there be suppuration and the necessity of lancing the
tonsil.

ENLARGED TONSILS.

The correct name for enlarged tonsils is hypertrophy
of the tonsils. The condition comes on after repeated
attacks of sore throat and acute tonsilitis, and in many
cases is dependent upon a scrofulous diathesis.

PATHOLOGY.

Following the repeated acute attacks there ensues a
sub-acute, or chronic inflammation of the glandular tissue of the tonsil, there is an exudation into the intercel-
lular spaces, which organizing, becomes hard, and forms
a firm hypertrophy, the lacunæ or crypts become diseased, and hold large masses of cheesy secretion, which,
in many cases, gives a foul odor to the breath.

TREATMENT.

In typical cases, and in those who have trouble winter
after winter the treatment is essentially operative.

Enlarged tonsils, of short duration, may be frequently
reduced by swabbing with glycerine three times a day,
and taking phytolacca and podophyllin internally.

A patient with enlarged tonsils is very liable to acute
exacerbations, and therefore should pay special attention
to hygienic treatment similar to that for catarrhal diseases.

When the glycerine and internal treatment fail an
operation should be resorted to. For this the tonsillotome is probably the best instrument. I prefer that of
Matthieu. (See Fig. 18.)

Fig. 13.

Swab the tonsil to be operated upon several times with a four per cent. solution of cocaine, then under good light, engage the tonsil in the ring of the tonsillotome and amputate. The tonsillotome is supplied with a small spear, which transfixes the tonsil before amputation, so that it does not drop into the throat.

It is seldom necessary to excise the entire tonsil clear to its base, but rather take that part which protudes inwards beyond the pillars of the fauces. In this case, the hemorrhage following is usually small, and can be controlled by gargling with ice water, or painting the bleeding surface with one of the iron salts.

The after treatment consists of avoiding exposure of any kind, and gargling with some antiseptic solution. I prefer the following after throat operations.

R—Calendula, 3 ss.
Water, O j
Sig.: Gargle frequently.

The prognosis in these cases is good, but it is well to remember that without operation the tonsil is absorbed, and gradually disappears after thirty years of age, a cure resulting spontaneously.

MOUTH BREATHING.

Mouth breathing occurs first as a result of habit, from hypertrophic rhinitis, from growths and polypi in the nose and from enlarged tonsils; remedying the several diseases will overcome the mouth breathing.

The consequences of mouth breathing are a stupid and silly facial expression, a dry throat, follicular pharyngitis, catarrhal laryngitis and in extreme cases a narrow or chicken-breasted chest and lung disease.

When the stenosis is overcome the patient must keep the mouth closed throughout the day, notwithstanding the amount of will power required. At night he must cover the mouth with a rubber-dam such as a dentist uses, tying it back of the neck. Another means is to arrange some kind of a bandage under the chin and over the head, thus causing the jaw to be closed tightly during sleep.

This is perhaps the best method, and even where disease and not habit is the cause, should always be practiced; also spraying the nose just before retiring with camphor-menthol is a great aid.

EXAMINATION OF THE LARYNX.

In making an examination of the larynx the same instruments are used as in making an examination of the post-nasal space, except in this case the tongue-depressor and rubber cord are not used. (See beginning of appendix.) The patient is seated in the same manner and draws his tongue out and downwards by grasping the end of it with his handkerchief. Use the largest size rhino-

Fig. 19.

scope, No. 4, and when warmed and in position the patient making the sounds "ah" and "eh" the vocal chords come in view and are plainly seen in the mirror of the rhinoscope. (See Fig. 19.)

The examination is comparatively easy; one point to be remembered is, in case of an irritable throat do not let the patient see that you have been unsuccessful, but try again at the next sitting.

DISEASES OF THE LARYNX.

ACUTE LARYNGITIS

Is a most dangerous form of disease. It usually commences with a slight chill, soreness and stiffness of the throat, difficulty of swallowing, and sense of constriction and desire to clear the throat. Following the chill, febrile reaction comes up, and is quite intense, considering the extent of the inflammation. Then a dull pain is felt in the throat, the sense of constriction is markedly increased, and there is tenderness on pressure; the voice is harsh, hoarse, or stridulous, and there is a frequent dry short cough. If the throat is now examined, the fauces will be found red and tumid, and when the tongue is pressed down the epiglottis may be seen erect, swollen and red. In the course of from twelve to twenty hours, the inflammation has markedly diminished the aperture of the glottis, the voice becomes small, piping, whispering, and soon suppressed. The breathing is difficult, inspiration being sibilous, shrill, prolonged and laborious, the larynx being forcibly drawn down on each attempt to inflate the lungs. The cough is stridulous and convulsive, and attended by attacks of spasm of the glottis, which threatens suffocation, the expectoration being scanty and viscid, and removed with difficulty. In the last stage of the disease, the patient exerts all his power in respiration, sitting upright and grasping objects in reach to bring into play the external inspiratory muscles. The countenance is pale and anxious, the lips livid, and the eyes almost start from their sockets, the extremities are cold, and covered with a clammy perspiration. Soon a low delirium or coma comes on, the pulse becomes more feeble and intermittent, imminent symptoms of asphyxia appear, and the patient rapidly sinks.

DIAGNOSIS.

The diagnosis is readily made in these cases, from the peculiar character of the voice, cough, location of soreness and constriction, and extreme difficulty of breathing; in asthenic laryngitis, by the marked difficulty of inspiration and freedom of expiration.

PROGNOSIS.

The prognosis is favorable in the first form, and even in the second, if the treatment is prompt and active, but doubtful in the third.

TREATMENT.

In the acute affection, means to cause relaxation of the larynx are of the utmost importance, giving us time to arrest the inflammation. For this purpose, we employ cloths wrung out of hot water, frequently changed, and the additional use of equal parts of oils of lobelia and stillingia, with just sufficient alcohol to cut them. Dry cups, or the cups and scarificator may be employed with marked advantage, if properly used. In addition inhalation of equal parts of vinegar and water, or either alone, is highly useful.

Internally, the most efficient remedies are the acetous tinctures of lobelia and sanguinaria, and syrup, equal parts, given in teaspoonful doses every five or ten minutes. It should be employed so as to keep up continuous nausea, but not to produce vomiting, unless it be found that such nausea does not produce the general relaxation necessary, when the compound powder of lobelia, in infusion, may be given so as to produce thorough and sufficiently continued emesis to accomplish the desired result.

Instead of this active treatment, we might rely upon the use of the small dose of aconite alone, as in croup— the stillingia liniment being the external application.

CHRONIC LARYNGITIS.

Chronic laryngitis may arise from an improperly treated catarrhal laryngitis, quite frequently from an extension of the chronic inflammation of the pharynx. Great and prolonged exercise of the voice, as in public speaking, singing, etc., is a prominent cause. Syphilis, also, not unfrequently affects the larynx, the disease being very persistent and intractable.

MINISTERS' SORE THROAT.

The first form of chronic laryngeal disease is designated as ministers' sore throat, and, as its name would indicate, is caused by the prolonged use of the vocal organs. It is not, however, confined to the ministry, but is met with in other public speakers and in singers, and less frequently among those who have had no special occasion for over-exercise of the organ.

This disease is not a true inflammation, at first, but rather an *irritable larynx;* the structures being in that condition that slight causes are sufficient to induce irritation and determination of blood. Continuing on, however, a true laryngitis is developed in time.

The first evidence of ministers' sore throat, is a sensation of irritation, with spasmodic contraction and cough on over-exertion of the vocal organs. As it progresses, this is more easily excited, and the person finds the voice becoming rough and harsh, and that he is losing control over it. At a further advanced stage, the voice is hoarse, at times sinking to a whisper, and the formation of words requires considerable effort; and, finally, speaking in ordinary conversation is difficult and unpleasant, and public speaking or singing impossible.

SYMPTOMS.

Chronic laryngitis usually comes on slowly and insiduously, the patient being hardly aware that he is suffering from a serious disease, until it is confirmed. The first symptoms are soreness of the throat when speaking, with constriction, slight alteration of the voice, cough, and expectoration, which comes on after slight exposure, or over-exertion of the larynx. These symptoms are ameliorated in a short time, and the patient thinks it is but a slight cold, from which he is recovering. As time advances, however, the attacks become more frequent, last longer and do not so nearly disappear. The disease being fully established, there is a constant, uneasy sensation in the region of the larynx, the voice is seriously altered, and there is a constantly annoying cough, with expectoration. The expectoration is at first scanty and mucus; but, as the disease advances, it is muco-puriform, sanious, concreted into lumps, or consists of almost pure pus. Hemorrhage occurs in the latter stages, sometime in very large quantity. If the throat is examined, we notice the evidence of chronic inflammation of the fauces, pharynx, epiglottis, and we reasonably suppose that the mucous membrane of the larynx corresponds in appearance; with the laryngoscope we are enabled to view the internal surface of the larynx, and determine its condition tolerably accurately.

A person suffering from "ministers' sore throat," or chronic laryngitis, is very subject to take cold, and thus every change in the weather, or slight exposure, is followed by an increase of the disease. A very important part of the treatment of every case, therefore, will be directed to obviate this.

The impairment of the general health is usually in direct ratio to the severity of the local affection. At the commencement, the patient complains simply of debility, with some failure of the digestive organs, and sometimes torpor of the secretions. When it has progressed for some months he is unable to attend to business; there is loss of flesh and strength; marked impairment of the digestive functions, and of excretion. Now, frequently the system becomes so depressed that tubercles are deposited in the lungs, the symptoms of phthisis are developed, and the disease runs a rapid course to a fatal termination.

DIAGNOSIS.

We diagnose chronic laryngitis by the unpleasant sensations in the region of the larynx, the cough and expectoration, the appearance of the throat, and the absence of physical signs of other diseases of the respiratory apparatus.

PROGNOSIS.

Ministers' sore throat can be readily cured in the majority of cases, if the person will give the vocal organs rest; usually from four to twelve months will be required.

The prognosis in confirmed laryngitis is not favorable, as but few have the patience necessary to persist in the use of remedies until a cure is affected   It can be cured, but it requires time and perseverance, otherwise the disease is as fatal as confirmed phthisis.

TREATMENT.

The treatment of ministers' sore throat is in part specific, and when the general health is but little impaired, but one or two remedies will be required. I prescribe in this case :

> R—Fluid extract of collinsonia.
> Simple sirup, *aa.* M.

A teaspoonful three or four times a day. If nutrition was somewhat impaired I would alternate with this:

> R—Tincture of nux vomica, gtt. x.
> Tincture of muriate of iron, ℨss.
> Glycerine, ℨ jss.
> Simple sirup, ℨ ij. M.

The patient is directed to use the cold vinegar pack to the throat on going to bed at night, using flannel cloths and wringing them quite dry ; washing the neck and shoulders with cold water in the morning, drying with brisk friction. There is no other means so certain to prevent the continued taking cold as this free bathing with cold water, and it should be insisted upon as an indispensable part of the treatment.

In the treatment of chronic laryngitis there are two prominent indications ; to relieve local irritation and give the larynx rest, and to improve general nutrition.

The patient must be impressed with the necessity of giving the organs rest, and hence, of so arranging his intercourse with others that there shall be no occasion for much talking. It is also well to show them that such conversation as is indispensable can be conducted in a low key with much greater comfort and with less exertion than any other. They should also understand clearly the necessity of controlling the cough by the influence of the will, for a cough is always a source of irritation, and must be kept in check by some means.

The remedies to relieve laryngeal irritation will vary in different cases. In cases where there is pharyngeal disease we will find that the use of those local remedies recommended for chronic pharyngitis will be among our most important means.

The gargle of Hamanelis is especially a favorite of mine, as it gives tone to the mucous structures, and relieves the tendency to stasis of blood. Stillingia is another excellent remedy in these cases, using it in the form of the compound tincture of the oils of stillingia, cajeput and lobelia; one drop on a lump of sugar every two or three hours, or a trituration of oil of stillingia with white sugar and gum arabic.

As a general rule, we will obtain the best results from remedies allowed to dissolve on the tongue and then swallowed slowly. The irritation seems to point in the fauces, and agents used in this way influence the parts from which the cough seems to arise. I will give a formula of each kind; narcotic, astringent and stimulant, and the practitioner will readily see how he can combine remedies to suit the particular case in hand:

R—Sulphate of morphia, gr. j.
    Chlorate of potash, ℥j.
    Powdered gum arabic.
    White sugar, aa. ℥ij. Divide in 20 parts.

10

R—Alum, 3 ss.
   Tincture of aconite, gtts. v.
   Powdered gum arabic.
   White sugar, *aa.* 3 ij.   Divide in 20 parts.

R—Capsicum, gr. v.
   Chlorate of potash 3 j.
   Gum arabic (powder), 3 ij.   Divide in 20 parts.

Where there is evidences of structural change in the larynx, the tissues being enfeebled, we will find the use of the spray apparatus of importance.   In this way we employ solutions of salicylic acid, permanganate of potash, sulphurous acid, iodine, and other remedies of like character.   It is well to combine a narcotic with these to prevent irritation and quiet cough.

But in a large majority of cases we will find the use of the vinegar pack, cold water sponging, and collinsonia, as named at first, will be all that is necessary, if the patient will give us his assistance and will persevere.

The *second* indication requires careful attention.   It may be stated, as a general rule, that no chronic inflammation can be arrested, or structural change repaired, unless there is good blood and active nutrition.   In chronic laryngitis there is a deterioration of the blood and an impaired nutrition which finally results in tuberculosis.   In severe cases it is the same in kind as the first, differing only in degree, hence the great importance of a restorative treatment.

Fortunately we are able to select our general remedies so that we may obtain a favorable local influence.   Thus the collinsonia, which relieves laryngeal irritation, is also an excellent tonic ; and a combination of muriated tincture of iron with glycerine is an excellent topical stimulant and demulcent, as well as one of the best forms of a restorative.

In some cases our patient will require the stronger tonics and restoratives; as the triple phosphate of quinia, strychnia and iron. Cod liver oil, when kindly received by the stomach, will answer an excellent purpose where the patient is thin in flesh, and when there is an exalted evening temperature. In those cases where the tongue is coated in the morning, with gaseous distension of the stomach after eating, and sometimes fœtid eructations, I frequently make the following prescription:

R—Podophyllin, grs. ij.
Hydrastine, grs. x.
Phosphate of soda, ℥ ij. M.

Triturate thoroughly and divide in twenty parts, and give one three times a day.

With regard to the use of nitrate of silver in chronic laryngitis I am satisfied that it has done far more harm than good. In some cases it can be applied to the fauces and pharynx as a topical stimulant with advantage, and associated with other treatment, a cure will result. But that its application within the larynx by a probang is good treatment in any case, I deny. I am well satisfied that in ninety-nine out of a hundred cases, the application is not made to the larynx but to the pharynx and œsophagus. But fortunately for the sufferers from laryngitis the *raid* on the larynx with probang, *a la* Dr. Horace Green, has passed by, and will shortly be recollected as one of the periodical absurdities of medicine.

INDEX.

C

S

T

DESCRIPTIVE CATALOGUE

—— AND ——

PRICE-LIST

—— OF ——

Eclectic ◦ Text-Books.

FOR SALE BY

John M. Scudder's Sons.

MEDICAL PUBLISHERS,

301 Plum Street,   *Cincinnati, Ohio.*

MAY 1, 1895.

Any Book in this List sent, post-paid, on receipt of Price.

AS a school of medicine we profess to have a distinctive practice, unlike either our old school or homœopathic neighbors. We claim to use different remedies, or in different form and dose, and for different effects. We boldly claim a more successful practice than either of our competitors, and this claim can only be based upon different principles, a different therapeutics, and a different materia medica.

We must, therefore, have distinctive books which clearly state *our* methods of practice. Old-school works will not serve this purpose, neither will homœopathic. With the pretensions we make, if we can not show that we have such works, and depend upon them, we are frauds of the first magnitude.

In the early days of Eclecticism, the need of text-books was clearly seen, and great sacrifices were made to furnish them. The writers toiled without pay, and to publish the earlier works they practiced the most rigid economy for years to command the money. By these means we had Beach's works, Jones and Morrow's Practice, King's Dispensatory, and some others. The making of books was not an easy nor a profitable job.

Now we have a full list of text-books, or books of reference, and by frequent revision they are kept fully up to our practice of to-day. They have been very successful, more so than any American books in the market, and this is the best evidence of their value. They are **bought by all schools** of medicine, and when bought they are brought into active use.

2

ECLECTIC TEXT-BOOKS.

BEACH, WOOSTER, M. D.
The Founder of Eclecticism.

The American Practice.

1 vol. 8vo; 873 pages; sheep. Price $4.50.

ELLINGWOOD, FINLEY, M. D.
Professor of Chemistry in Bennett College of Eclectic Medicine.

Annual of Eclectic Medicine and Surgery.

1890, Vol. 1. 8vo; 337 pages; cloth. Price $2 00.
1891, Vol. II. 8vo; 425 pages; cloth. Price $2.35.

GOSS, I. J. M., M. D.
Professor of Practice of Medicine in the Georgia College of Eclectic Medicine and Surgery.

Materia Medica, Pharmacology, and Special Therapeutics.

8vo, 536 pages; cloth, price $3.50; sheep, $4.50.

The Practice of Medicine; or, The Specific Art of Healing.

8vo, 569 pages; cloth, price $3 50; sheep, $4.50.

HOWE, A. JACKSON, M. D.
Late Professor of Surgery in the Eclectic Medical Institute, Cincinnati.
"Prof. Howe was recognized as one of the ablest teachers in this country, and an operating surgeon with but few peers in the West."

The Art and Science of Surgery.

Revised edition. 8vo, 886 pages; sheep. Price $7 00.

Diagnosis and Treatment of Dislocations and Fractures

Fourth edition. 8vo, 426 pages; sheep. Price $4.00.

Operative Gynæcology.

8vo, 360 pages; sheep. Price $4 00.

JEANCON, J. A., M. D.
Professor of Pathology in the Eclectic Medical Institute, Cincinnati.

Pathological Anatomy and Physical Diagnosis.

Royal folio, 100 full page Colored Illustrations, 100 pages of text; half morocco. Price $20.00.

Diseases of the Sexual Organs. (MALE AND FEMALE.)

Royal folio, 80 full-page Colored Illustrations, 160 pages of text; half morocco. Price $20.00.

KING, JOHN, M. D.
Late Professor of Obstetrics in the Eclectic Medical Institute, Cincinnati.
"Prof. John King, who was a teacher for more than half a century, is too well known to require more than a catalogue of his books." They are as follows:

The American Dispensatory. With Supplement by J. U. LLOYD.

Seventeenth edition. 8vo, 1,611 pages; sheep. Price $10.00.

Diagnosis and Treatment of Chronic Diseases.

The New American Family Physician.

8vo, 1,042 pages; morocco. Price $6.50.

The American Eclectic Obstetrics.

Revised, re-written and enlarged, by R. C. WINTERMUTE, M. D.
Ninth edition. 8vo, 757 pages; sheep. Price $6.50.

Woman, Her Diseases and Their Treatment.

Fourth Edition. 8vo, 366 pages; sheep. Price $3.50.

Urological Dictionary.

8vo, 266 pages; sheep. Price $2.00.

LLOYD, J. U.

Professor of Chemistry and Pharmacy in the Eclectic Medical Institute. Cincinnati; Vice
President of the American Chemical Society; Associate Author of the Supple-
ment to the American Dispensatory; Author of the Pharmacy and
Chemistry of the Student's Pocket Medical Lexicon.

The Chemistry of Medicines.

A text and reference book for the use of students, physicians and pharmacists; embod-
ying the principles of chemical philosophy, and their application to those chemicals that
are used in medicine and in pharmacy, including all those that are officinal in the Pharmacopœia of the United States.

One volume, large 12mo, 451 pages. Cloth, $2.75; leather, $3.25.

LOCKE, FREDERICK J., M. D.

Dean and Professor of Materia Medica and Therapeutics in the Eclectic Medical Institute,
Cincinnati, Ohio.

HARVEY W. FELTER, M. D., Collaborator,

Demonstrator of Anatomy and Quiz-master in Chemistry in same.

Syllabus of Eclectic Materia Medica and Therapeutics.

12mo, 450 pages; cloth. Price $2.50.

MERRELL, ALBERT, M. D.

Formerly Professor of Chemistry and Pharmacy in American Medical College, St. Louis.

Digest of Materia Medica and Pharmacy.

8vo, 512 pages; cloth. Price $4.00.

McMILLEN, BISHOP, M. D.

Of Columbus, Ohio.

Mental and Reflex Diseases.

12mo, about 300 pages; cloth. Price $2.00 net.
(In preparation.)

NIEDERKORN, JOSEPH S., M. D.

A Ready Guide to Specific Medication.

16mo, limp cloth. Price $1.00.

SCUDDER, JOHN M., M. D.

Late Professor of the Practice of Medicine in the Eclectic Medical Institute, Cincinnati.

The Eclectic Practice of Medicine.

Fourteenth edition, revised. 8vo, 816 pages; sheep. Price $7.00, post-paid. $6 00, post paid, to Journal subscribers.

The best recommendation of this work comes in the statement. **Fourteenth Edition.** It is *the* authority of our school of medicine, and thousands of sick are daily treated according t) it. Thus far it has proven sufficient, and has given a success that others have failed to obtain.

(SCUDDER, J. M., CONTINUED.)

The Principles of Medicine.

Fifth edition. 8vo, 350 pages; sheep. Price, post-paid, $4.00.

This is a study of the elements of disease and the principles of cure. It is the basis of our practice, and, as we think, the practice of the future. It gives a rational basis for medical practice.

The Eclectic Practice in Diseases of Children.

Sixth edition. 8vo, 486 pages; sheep. Price, post-paid, $5.00.

If there is one thing more than another that we take pride in, it is our success in the treatment of children. The teaching of pleasant remedies, in small doses, for direct effect has relieved thousands of children from the horrors of "regular" medicine.

A Practical Treatise on the Diseases of Women.

ILLUSTRATED BY COLORED PLATES AND NUMEROUS WOOD ENGRAVINGS.

WITH A PAPER ON DISEASES OF THE BREASTS.

Fifteenth edition, revised. 8vo, 534 pages; sheep. Price, post-paid, $4.00.

This work has stood the test of twenty years, and, as revised, gives our treatment of to day.

Specific Medication and Specific Medicines.

Fourteenth edition, fourth revision. 12mo, 432 pages; cloth, $2.50.

Specific Diagnosis.

Ninth edition. 12mo, 388 pages; cloth, $2 50.

These companion volumes have had a larger sale than any other medical works in this country. They appeal to the feeling every thinking physician cherishes, that there must be something certain in medicine, if it can be discovered. They have had a very marked influence upon medical practice, not only of our own school, but also on regular medicine and homœopathy.

The American Eclectic Materia Medica and Therapeutics.

Tenth edition. 8vo, 748 pages; sheep. Price, post-paid, $6 00.

The Eclectic Practice of Medicine for Families.

Twenty-first edition. Cloth, $3 00; sheep, $4.00; half morocco, $5.00.

This work contains all of medicine that a family should know. It is anatomy, Physiology, Hygiene, Practice, Materia Medica, Surgery, and Obstetrics. It is concise, plain and correct, and will not lend to household drugging. Liberal offers to agents. Write for terms.

On the Reproductive Organs and the Venereal.
With Colored Illustrations of Syphilis.

Second edition. 8vo, 393 pages; sheep. Price, post-paid, $5.00.

Medicated Inhalations.

Fourth edition, revised, with new Appendix, by WM. BYRD SCUDDER, M.D

8vo, about 125 pages; cloth. $1.00 net.

STEVENS, J. V., M. D.

Professor of Diseases of Children in the Bennett College of Eclectic Medicine and Surgery

Annual of Eclectic Medicine and Surgery.

1892, Vol. III. 8vo, 444 pages; cloth. Price $3.00.
1893, Vol. IV. 8vo, 503 pages; cloth. Price $3.00.
1894, Vol. V. In preparation; cloth. Price $3.20.

Successive volumes issued in each November.

WEBSTER, HERBERT T., M. D.

Professor of the Principles of Medicine and Pathology in the California Eclectic Medical College, San Francisco.

The Principles of Medicine.

8vo, 163 pages. Cloth, $1.50.

Dynamical Therapeutics.

A work devoted to the Theory and Practice of Specific Medication, with special reference to the newer remedies.

8vo, 853 pages. Cloth, $5.00; sheep, $6.00.

WINTERMUTE, ROBERT C., M. D.

Professor of Obstetrics and Diseases of Women and Children in the Eclectic Medical Institute, Cincinnati.

American Eclectic Obstetrics.

A new edition of the standard edition of King's. Thoroughly revised and re-written.

8vo, 757 pages; sheep. Price $6.50.

WATKINS, LYMAN, M. D.

Professor of Physiology in the Eclectic Medical Institute; Physician to the Eclectic Hospital of Cincinnati, etc.

A Compendium of the Practice of Medicine.

12mo, cloth; 400 pages. Price $2.50 net.

(In preparation)

ECLECTIC MEDICAL JOURNAL.

JOHN K. SCUDDER, M. D.

MANAGING EDITOR.

48 to 64 pages Monthly.   *$2.00 per Annum in advance.*

FIFTY-FIFTH YEAR.

The acknowledged organ of liberal medicine, and a strong advocate of the doctrines of Specific Medication.

A Ready Guide to the Study of Specific Medication.

BY JOSEPH S. NIEDERKORN, M. D.

16mo, 116 pages; limp cloth.   Price $1.00.

NOTICES OF THE PRESS.

We are glad to see so excellent a little work in this special line, The book does not treat specific diseases, but considers organs or functions, and considers the specific action of our remedies upon them. The book is a little gem, and is worth many times its cost in putting our knowledge of the action of remedies into a tangible and accessible form. We trust every reader of the *Times* will purchase a copy of the book at once. —*Chicago Medical Times.*

KING'S ECLECTIC OBSTETRICS,

REVISED, REWRITTEN, AND ENLARGED, BY

ROBERT C. WINTERMUTE, M. D.

Professor of Obstetrics and Diseases of Women and Children in Eclectic Medical Institute

Ninth edition. 8vo, 757 pages; sheep. Price $6.50.

WHAT OTHERS SAY OF THE WORK.

For the past thirty-five years, King's *Eclectic Obstetrics* has been the standard text-book on obstetrics for practitioners and students of the Eclectic school of medicine. The fact that for the last fifteen years there has been no new edition of this work is sufficient to demonstrate the great need of this revision, which has been undertaken by Dr. Wintermute. The additions and revisions which have had to be made in every chapter of the work have necessarily been great, and Dr. Wintermute is to be congratulated upon the thoroughness with which he has done the work.—*Cincinnati Lancet-Clinic.*

It will be extremely gratifying to all students of Eclectic medicine to know that this most excellent work has been fully revised and brought up to date. The labors of Dr. Wintermute are apparent in the absolute modern therapeutics. But little improvement was possible on the obstetrical methods of Dr. King. When the book was written it was twenty years ahead of its time, or more. Those of us who have won laurels for ourselves by following its methods do not need the recommendation of others to convince us of its value. But therapeutics has advanced greatly in a few years, and in this particular the book needed the thorough revision Dr. Wintermute has given it. It is now the most advanced and practical work of its size published, and we bespeak for it an immense sale.—*Chicago Medical Times.*

DYNAMICAL THERAPEUTICS.

A WORK DEVOTED TO THE

Theory and Practice of Specific Medication,

WITH SPECIAL REFERENCE TO THE NEWER REMEDIES ;

With a Clinical Index, adapting it to the needs of the busy practitioner.

BY HERBERT T. WEBSTER, M. D.

Professor of the Principles of Medicine and Pathology in the California Medical
College (Eclectic), San Francisco.

8vo, 680 pages.   Cloth, $5.00 ; sheep, $6.00.

WHAT OTHERS SAY OF THE BOOK.

In perusing this book, the reader is first impressed with its exhaustive
thoroughness, and next with the author's scholarship and modesty.
Prof. Webster has certainly been a tireless student and investigator, for
every line of his work discovers the most profound knowledge of the
topics treated, and these cover all the essentials of practical therapeutics.
High as our opinion of Dr. Webster's ability has always been, his book
gave us a most grateful surprise.   There is nothing in Eclectic literature
superior to it, and only a little which can be said to be equal to it.   It is
restful, satisfying reading, because of its reliable solidity, and its inno-
cence of egotism and ostentation.   *   *   *   A commendable
feature is the insertion of papers on special subjects by our brightest
men.   Thus, he utilizes "Notes on Pharmacy," by that highest authority,
Prof. J. U. Lloyd, and there are numerous other quotations from men
who are bright lights in their specialties.   Altogether, Prof. Webster has
made an addition to the Eclectic book world which is of inestimable
value, and which will place him in the van of the world's benefactors.—
Eclectic Medical Gleaner.

We are glad to bring this work to the notice of our readers.   It is a
new and very practical study of Specific Medication by one who handles
remedies with skill, and has shown more than usual capacity to teach.
The first study is the "Principles of Selection," and embraces 100 pages.
It endeavors to show the reader that there are sound principles or reasons for the selection of remedies, and that when thus selected with
care the specific effects may confidently be expected.   The successful
physician trains himself to correct observation.   He sees the varying
expressions of disease, recognizes their relation to the known action of
remedies, and makes a rational adaptation of the one to the other.

Part II. commences the study of remedies under the head of "Specific
Therapeutics."   The first consideration is the tissue remedies of Schuessler, from which the author seems to have obtained better results than I
have.   As we "prove all things, and hold fast that which is good," the
reader may wish to try the remedies, and here is a guide without buying
Schuessler.

The remedies are further taken up under the headings of Blood Mak-
ers; Antiseptics; Antizymotics; Correctives; the Periosteum; the Ar-
ticulations; the Nervous System; the Circulatory System; the Lymphatic System ; the Ductless Glands; the Digestive Organs; the Respiratory Organs; the Urinary Organs; the Sexual Organs; the Muscles; the
Skin ; the Eye ; the Ear.

This classification may seem familiar to you, if you studied medicine
in Cincinnati, or have been a reader of the JOURNAL.   But I assure you
Dr. Webster has made an original study under these heads, and whilst
he is willing to admit indebtedness to his old preceptors, he claims to be
heard for his own work.   I cordially recommend this book to my readers as one to be read and studied.—*Eclectic Medical Journal.*

TEXT-BOOKS,

Other than Eclectic, in use by the Students of the Several Eclectic Colleges.

	Binding.	Publishers' List Prices.
Bell's Comparative Anatomy and Physiology, -	Cloth,	$2.00
Brubaker's Physiology, - - - - -	"	1.00
Cleveland's Lexicon, - - - - -	"	75
" " - - - -	Leather,	1.00
Dunglison's Dictionary, - - - -	Sheep,	8.00
Gould's Dictionary, - - - - -	Cloth,	3.25
" " Indexed, - -	Half Mor.	4.25
Gray's Anatomy, Plain, - - - -	Cloth,	6.00
" " " - - - -	Sheep,	7.00
" " Colored Plates, - - -	"	8.00
Huxley's Physiology, - - - -	Cloth,	1.75
Huxley & Martin's Biology, - - - -	"	2.50
Kilgour's Indications, - - - - -	"	1.00
Kirke's Physiology, - - - - -	"	4.00
" " - - - - -	Sheep,	5.00
Klein's Histology, - - - - -	Cloth,	1.75
Lloyd's Elixirs, - - - - -	"	1.25
Nancrede's Anatomy—C. P., - - -	Oil Cloth,	2.00
Nettleship on the Eye, - . - -	Cloth,	2.00
Norton's Ophthalmic Therapeutics, - - -	"	2.50
Norton's Ophthalmic Diseases, - - - -	"	3.50
Ranney's Nervous Diseases, - - -	"	5.50
Reese's Medical Jurisprudence, - - -	"	3.00
Rohe's Hygiene, - - - - - -	"	2.50
Roosa on Diseases of the Ear, - - - -	"	5.50
Hyde's Diseases of the Skin, - - - -	"	5.50
Sajous on the Nose and Throat, - - - -	"	4.00
Steele's Physics, - - - - -	"	1.17
Tenney's Natural History, - - - -	"	1.38
Thomas' Dictionary, - - - - -	"	3.00
" " - - - - -	Sheep,	3.50

www.ingramcontent.com/pod-product-compliance
Lightning Source LLC
Chambersburg PA
CBHW021808190326
41518CB00007B/500